SharePoint Architect's Planning Guide

Create reusable architecture and governance to support
collaboration with SharePoint and Microsoft 365

Patrick Tucker

BIRMINGHAM—MUMBAI

SharePoint Architect's Planning Guide

Group Product Manager: Alok Dhuri
Publishing Product Manager: Shweta Bairoliya
Senior Editor: Nithya Sadanandan
Technical Editor: Maran Fernandes
Copy Editor: Safis Editing
Project Coordinator: Manisha Singh
Proofreader: Safis Editing
Indexer: Manju Arasan
Production Designer: Ponraj Dhandapani
Marketing Coordinator: Deepak Kumar and Rayyan Khan
Business Development Executive: Uzma Sheerin

First published: August 2022

Production reference: 1190822

Published by Packt Publishing Ltd.
Livery Place
35 Livery Street
Birmingham
B3 2PB, UK.

ISBN 978-1-80324-936-0

www.packt.com

With credit to all those over the years who've been on this journey with me. Thanks to Paul Stork for your willingness to be the technical reviewer and for being a great mentor and role model. I appreciate my dear friends at KiZAN and the new team at Microsoft. Thanks to Manisha, Shweta, Nithya, Uzma, and the Packt team for your patience and expertise. Thanks to my parents and brother for your love. This is dedicated to Merky and Sullivan and my family for always giving me your best and forgetting my worst.

- Patrick Tucker

Contributors

About the author

Patrick Tucker has brought together a passion for collaboration technologies and helping people to best experience the benefits of those tools for almost 25 years. He has worked as a trusted advisor with hundreds of organizations of all sizes as a consultant, speaker, Microsoft Certified Trainer, change manager, solution and information architect, business analyst, project manager, software developer, writer, and leader.

Patrick has spoken at SPTechCon and SharePoint Fest, as well as numerous SharePoint Saturdays and other Microsoft 365 community events. Today, he is a cloud solution architect, with a focus on adoption and change management in the Culture and Cloud Experience practice at Microsoft. He lives in the Louisville, Kentucky area and blogs at www.tuckersnet.com.

About the reviewer

Paul Papanek Stork is the owner/principal architect at Don't Pa..Panic Consulting. He has been working in the information technology industry for over 35 years and played key roles in several enterprise SharePoint architectural design reviews, intranet deployment, application development, and migration projects. He has been awarded the **Microsoft Most Valuable Professional (MVP)** award consecutively for the last 14 years and earned his **Microsoft Certified SharePoint Master's (MCSM)** certification in 2013.

His core skills include architecting and customizing SharePoint environments both on-premises and in the cloud. He has significant experience leading SharePoint migration teams, troubleshooting infrastructure issues, and evaluating SharePoint environments for adherence to "best practices" and industry standards. His current focus is on Office 365 security, SharePoint Online, and Microsoft Power Apps/Power Automate. He's authored several books and is a frequent contributor to the Microsoft Power Platform forums.

Table of Contents

3

Modern Options for Customizing SharePoint Online

Part 2: From Lone Wolf to Pack Leader – SPO Integrations with M365

4

Understanding M365 Groups as the Foundation of Collaboration

Part 3: From Tall to Flat – SPO Information Architecture

7

Up with Hubs, Down with Subs – Planning Hub Sites

8

The Mega World of Metadata

9

Keeping Things Secure – Permissions, Sharing, and DLP

Part 4: From Current to Change

10

The Human Side of SharePoint – Adoption and Managing Change

Index

Other Books You May Enjoy

Preface

SharePoint is no longer a standalone tool; in the Microsoft cloud, it is the backbone of collaboration for tools such as Teams and Yammer. It is also intertwined with M365 Groups. It works for and with the Power Platform, Microsoft Forms, and Microsoft Purview. This book is not a how-to document or a step-by-step guide. There are tons of online resources that give you that and help you better keep up with the pace of change. This book is a planning guide to help you with context, capabilities, and considerations for implementing SharePoint Online in the most successful way possible and helping your organization adopt it to meet its business needs.

Who this book is for

This book is for the long-time SharePoint on-premises admin who is making a leap to the cloud. It is for the IT architect who has experience in other areas but has never worked with SharePoint. While someone with a background in IT could use this book to learn about SharePoint Online, it is best used by those with some experience in the SharePoint world or of other aspects of Microsoft 365.

What this book covers

Chapter 1, *Classic versus Modern SharePoint*, will explore how the world of SharePoint changes as we move from on-prem farms to SharePoint Online. This will include a review of classic site templates, pages, and web parts, and how they have been either replaced or deprecated in the move to the modern experience. We will also use this chapter to introduce the capabilities of modern SharePoint sites and pages to readers who may not be familiar with them.

Chapter 2, *Making the Move – Migration Options and Considerations*, will look at how on-premises SharePoint can be moved to SharePoint Online and the common roadblocks along the way. We'll look at migration tools and understand the process of moving to the cloud, including the best practices for handling different types of content and customizations.

Chapter 3, *Modern Options for Customizing SharePoint Online*, will define the best practices for customizing the SharePoint Online UI, including site templates and site themes. JSON formatting for lists and adaptive cards will be reviewed along with options and best practices for SharePoint development using SPFx.

Chapter 4, *Understanding M365 Groups as the Foundation of Collaboration*, will look at how M365 Groups are the building blocks of team sites in SharePoint as well as Microsoft Teams, Yammer, and Outlook Groups. The provisioning and governance of Groups, including securing and labeling them, will be a major portion of this chapter.

Chapter 5, Integrating SharePoint Online and Other Collaboration Tools, will reinforce the point that SharePoint doesn't stand alone, but is a tool in the broader M365 toolbox. We'll explore how SharePoint is the core of tools such as Teams, Yammer, Lists, and Stream, but also works with Planner, To Do, and Microsoft Forms to form a complete solution.

Chapter 6, Making SharePoint More POWERful, will explore how SharePoint integrates with Power Automate for workflow management and processing, Power Apps for form customizations, and Power BI for visualizing data. We'll also plan for Power Platform governance.

Chapter 7, Up with Hubs, Down with Subs – Planning Hub Sites, will plumb the depths of the concept of hub sites available only in SharePoint Online. Hubs are a way to connect sites together to organize and navigate complicated organizational relationships. Viva Connections as an outcome of the work with hubs will also be explored.

Chapter 8, The Mega World of Metadata, will guide us along our planning journey to understand and best utilize metadata to drive the searching, sorting, and filtering of data in SharePoint and help us to leverage retention and sensitivity labels. SharePoint Syntex and Viva Topics will also be explored as knowledge management tools.

Chapter 9, Keeping Things Secure – Permissions, Sharing, and DLP, will deep-dive into how SharePoint security works in the cloud, including how M365 Groups play a part. Best practices for configuring and securing external sharing in SharePoint and OneDrive for Business, with the help of Microsoft Purview and DLP, will also be explored.

Chapter 10, The Human Side of SharePoint – Adoption and Managing Change, will focus on the adoption and change management aspects of moving your organization to SharePoint Online and M365. Our focus will be on the ProSci methodology of change management and practical advice for attaining deep and habitual adoption of our collaboration toolbox.

To get the most out of this book

Previous experience with SharePoint will be helpful to get the most out of this guide, as will any experience with the Microsoft 365 suite of tools. A desire to plan for, design, and architect compelling and adoptable solutions based on SharePoint is a must!

Software/hardware covered in the book	Operating system requirements
M365 tenant with SharePoint Online license	Windows, macOS, or Linux
Power Platform license	
SharePoint Online Admin Shell	

Questions at the end of each chapter are provided to guide your discovery and planning process. Please use the questions as a starting point to build your own architecture and governance document for SharePoint Online and the associated technologies that work alongside it in Microsoft 365.

Download the color images

We also provide a PDF file that has color images of the screenshots and diagrams used in this book. You can download it here: `https://packt.link/8fwP6`.

> **Important note**
> The content inside some images in the book may appear blurred. These are only the instances where the content is irrelevant, and the reader is supposed to see how the interface looks.

Conventions used

There are a number of text conventions used throughout this book.

`Code in text`: Indicates code words in text, database table names, folder names, filenames, file extensions, pathnames, dummy URLs, user input, and Twitter handles. Here is an example: "The path is `Apps/Yammer` and gets automatically created and connected when the community is built."

Bold: Indicates a new term, an important word, or words that you see onscreen. For instance, words in menus or dialog boxes appear in **bold**. Here is an example: "Each channel has a **Files** tab, which displays the set of files stored for that channel."

> **Tips or important notes**
> Appear like this.

Get in touch

Feedback from our readers is always welcome.

General feedback: If you have questions about any aspect of this book, email us at `customercare@packtpub.com` and mention the book title in the subject of your message.

Errata: Although we have taken every care to ensure the accuracy of our content, mistakes do happen. If you have found a mistake in this book, we would be grateful if you would report this to us. Please visit `www.packtpub.com/support/errata` and fill in the form.

Piracy: If you come across any illegal copies of our works in any form on the internet, we would be grateful if you would provide us with the location address or website name. Please contact us at `copyright@packt.com` with a link to the material.

If you are interested in becoming an author: If there is a topic that you have expertise in and you are interested in either writing or contributing to a book, please visit authors.packtpub.com.

Share Your Thoughts

Once you've read *SharePoint Architect's Planning Guide*, we'd love to hear your thoughts! Scan the QR code below to go straight to the Amazon review page for this book and share your feedback.

https://packt.link/r/1803249366

Your review is important to us and the tech community and will help us make sure we're delivering excellent quality content.

Part 1: From Farm to Cloud

This section sets the stage for SharePoint architects that have been on-prem farm admins and are now moving to M365. The goal is to present how classic and modern SharePoint differ; what the best practices are for theming and customization since we've lost Content Editor, Script Editor, and custom master pages; and the options, considerations, and best practices around migrating from on-prem to the cloud.

The following chapters are included in this part:

- *Chapter 1, Classic versus Modern SharePoint*
- *Chapter 2, Making the Move – Migration Options and Considerations*
- *Chapter 3, Modern Options for Customizing SharePoint Online*

1

Classic versus Modern SharePoint

There is an idiom that states *the more things change, the more they stay the same.* In the world of SharePoint, many familiar concepts have existed for a long time, but there have also been many changes over the years. With SharePoint Online, those changes are occurring at a rapid pace.

It's been a privilege to see SharePoint change and mature over the years, from its earliest days as an *extension* to the on-premises Office Online Server to becoming a *server-based product* of its own to becoming a *set of services in the cloud* that forms the backbone of file sharing and collaboration.

This chapter is designed for the person who has been a part of that history and worked with on-premises SharePoint but is new to the world of **Microsoft 365** (**M365**) and SharePoint Online. Maybe you've been an architect, an admin, a developer, a trainer, or a frustrated user. Cool! I've been there too. As with all technical books, this one sits along a point in that timeline, with a view of the past and a fleeting moment to be current at least or future-facing at best.

On the SharePoint timeline sits a pivotal divide. It is kind of like BC/AD or BCE/CE. We'll call it C/M – the divide between classic and modern SharePoint. In many ways, this divide can be seen as the divide between **SharePoint on-premises** versus **SharePoint in the cloud**. The dividing line is not quite that crisp, however. It is on that blurry line on which our first planning exercise begins, as we explore the following topics together:

- Modern building blocks
- The classics
- A mixed skyline
- The paths to modern
- Hybrid workloads
- Additional features
- IT governance
- Planning document

By the end of the chapter, we will have reviewed the five core areas most impactful to users of SharePoint Online. We will have looked at modern sites and web parts, ways to get from classic to modern, and the ability to combine the worlds of SharePoint Server and Office 365 by using hybrid mode.

Understanding modern building blocks

The term *modern* in this context usually leads us to one of two places. It could simply mean that we are now using **SharePoint Online** (and the default sites, pages, templates, and other services it provides) rather than using **SharePoint Server (SP Server)** on-premises. It most likely means that we are using the modern UI components introduced in SharePoint Online and made available to SP Server 2019 as well.

The modern UI components can best be defined as the collection of responsive, client-side UI elements and the structures that support them. These elements provide us with the building blocks of mobile-ready sites that automatically adjust to the screen and device they're on. The modern UI also provides a tremendously simple editing canvas that should enable both IT and content owners with little technical experience to quickly create and modify the content in a compelling and dynamic way.

There are five core areas where we can see this modern structure shining through:

- **Sites** (formerly known as site collections)
- **Lists and libraries**
- **Pages**
- **Web parts**
- **Navigation**

We will take a look at each of these structures in the following sections. Each section will dig deeper into one of the components that comprise modern SharePoint. I like to think of it as a nesting doll with one component containing another. Sites are the big doll with lists, libraries, and pages inside. Pages then contain web parts that we can configure to build our modern UI.

Sites

In SharePoint on-premises, we had a variety of site templates to choose from when provisioning a new site. Most sites were created as team sites, but we could get more specialized as the need arose. While there was a long list of site templates to choose from, they primarily differed in which features were activated by default. In SharePoint Server 2016, there were just over 50 site templates that would be returned when executing the `Get-SPOWebTemplate` cmdlet. You can see the full list here: `https://social.technet.microsoft.com/wiki/contents/articles/52674.sharepoint-server-2019-list-of-all-templates.aspx`.

For the most part, we could really just use the blank site template and enable the features we need at the site collection or site level. For example, if you wanted the capabilities of a **Document Center** site, you could start with a blank site template and add the **Document ID** and **Content Organizer** features.

The story is greatly simplified in SharePoint Online. While admins can create classic sites from most of the legacy templates, there are only two modern site templates – **Communication site** and **Team site**. Something to remember is that once a site is created from a template, we cannot change that template. Our only hope would be starting over and migrating data, so we must choose wisely. This is the choice we're given as a user creating sites or the defaults in the admin center:

Create a site
Choose the type of site you would like to create or learn more about <u>team sites</u> and <u>communication sites</u>.

Team site
Create a private space to collaborate with your team.

- Track and stay updated on project status
- Share team resources and co-author content
- All site owners and members publish site content

Communication site
Share information that engages and informs viewers.

- Create portals or subject-specific sites
- Engage dozens or thousands of viewers
- Few content authors and many site visitors

Figure 1.1 – Team site and Communication site template options

For admins, team sites can come in two flavors, depending on whether you want to connect the site to M365 Groups (which we'll explore in *Chapter 4*). Generally, while features are still present to an admin or owner, I have very rarely bothered with feature activation in the modern SharePoint world. Generally, it is no longer needed to make modern sites behave in specific ways. The out-of-the-box configuration is sufficient in most cases.

Lists and libraries

The pages automatically created for viewing lists and libraries will be modern in SharePoint Online by default. You can change this at the site level with the activation of a feature (to be consistent with my previous statement, I've never had to do this) called **SharePoint Lists and Libraries experience**. Activating the feature turns off the modern UI for list pages. You can also use the Advanced list or library settings to change the default list experience on a one-by-one basis. The following figure shows that option:

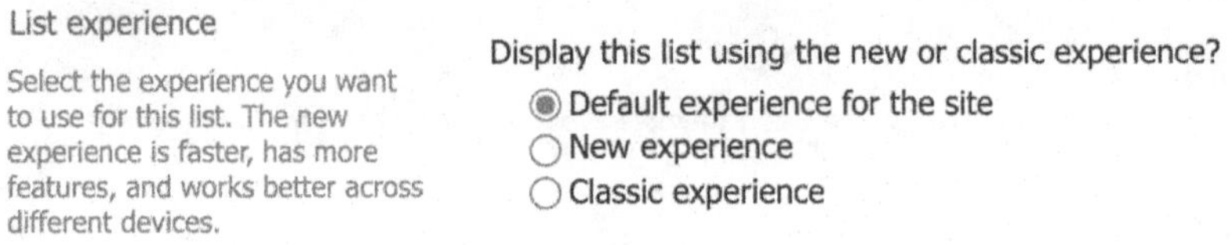

List experience

Select the experience you want to use for this list. The new experience is faster, has more features, and works better across different devices.

Display this list using the new or classic experience?
- ◉ Default experience for the site
- ◯ New experience
- ◯ Classic experience

Figure 1.2 – List experience on the List settings page

While it is possible to make that change for a particular list or even for all lists on a particular site, I wouldn't recommend it unless you have a compelling reason to do so. These modern pages are crucial to connect our SharePoint sites to features outside the platform (such as Power Apps and Power Automate) and to take advantage of the latest and greatest from Microsoft. The modern UI for system-generated list and library pages (for everything but the calendar that is) no longer features the classic ribbon, but a new streamlined UI filled with modern features, as seen in the following figure:

Figure 1.3 – The menu seen on list and library pages in SharePoint Online

Modern views are also much friendlier to work with and change. We can do so directly on the same page as the list, rather than having to navigate to a separate screen. Therefore, views can be manipulated in place. We can then save them as public or private. We will explore these options further in *Chapter 5, Magic Tool in the Toolbox – Integrating SPO with Other Collaboration Tools*, when we discuss Microsoft Lists as a standalone service. Lists brings new formatting and other enhancements to a single new app experience where users can interact with list data across sites they have access to as well as their own private lists created in OneDrive for Business.

Pages

For pages that we create, including the home page of the site, we have a great leap forward in ease of use and depth of functionality with modern UI enhancements. We have an entirely new canvas providing easier editing and updating. Modern pages also render correctly whether we're on a phone, a tablet, or viewing the page on a 49-inch wide-screen monitor. This works because Microsoft has built the underlying page structure to be responsive.

Modern pages are possible because each new modern site contains an enabled **site pages** feature that gives us a new **Site Page** content type. A modern page can include multiple sections with different layouts, without the use of the Publishing Infrastructure features or **page templates**. Moreover, the layout of the section can change as we build our page, so we could have a single, wide panel at the top with a three-column layout just underneath it, and back to a page width section. Section layouts can be changed even after the page is saved and web parts will adjust to fit into their container if they are moved.

On the following page, for example, we have a vertical section down the right side. A hero layout of four images is in a single-column section across the top, and a two-column section with some text and previews/links to videos is just below it. If we moved the hero images to the smaller section below, it would render as a slider, showing one image at a time. The following figure shows a hero layout rendered on a desktop monitor, displaying all four images at once:

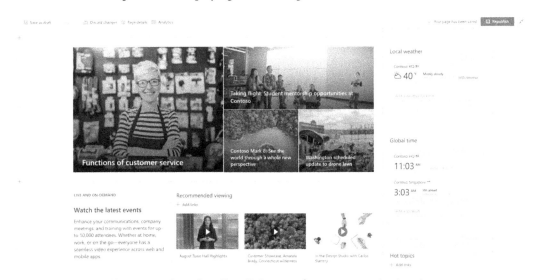

Figure 1.4 – A modern SharePoint page from a communication site

Microsoft has created and continues to curate a tremendously valuable resource to make sure we understand and get the most benefit from modern sites. The **SharePoint look book** contains loads of ideas, documentation, and visual samples, and you can deploy entire sites to your tenant directly from the site as well. The look book is available at `https://lookbook.microsoft.com/`.

Web parts

Classic pages are built with classic web parts. Modern pages are composed of modern web parts. That may sound elementary, but it is crucial to understand that we cannot mix and match components between the two worlds. They're both building blocks, but one is like LEGO bricks while the other is like Mega Bloks. They simply don't fit together.

Modern web parts are responsive, just like the pages that host them. When a page is placed into edit mode, so are all of the web parts, so there are fewer clicks for our content owners. The **Edit** panel that opens for each web part is also simpler and more consistent with the rest of the modern UI. Just look for a pencil icon to click and the details panel will open on the right:

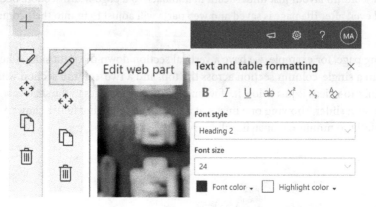

Figure 1.5 – Editing a modern web part

The look book noted earlier does a great job of indicating which web parts have been used with each template's home page, with hyperlinks to the Microsoft documentation to explain how to use each one.

Navigation

In both classic and modern SharePoint, we have navigation elements that can be hosted in one of two locations on the page. Team sites still provide a left-hand quick launch, while site navigation can still be found at the top of the page. Adding and editing links becomes more streamlined with the owner's ability to edit the navigation in place and use the **Change the Look** menu to decide on a more traditional **Cascading** look or the more updated **Mega menu**:

Figure 1.6 – An example of Mega menu

The biggest change to navigation in the modern world is through the use of hub sites, which we will explore in much greater depth in *Chapter 7, Up with Hubs, Down with Subs – Planning Hub Sites*. A hub provides a way for one site to be the logical parent of other sites. One advantage of this functionality is to provide global navigation that is at the very top of the page, above the site's navigation, which is consistent across all sites that are connected to a given hub. We can also use that same global navigation to supply links to the SharePoint Online app bar. The app bar is designed to be sticky to the left-hand side of all SharePoint Online sites and to also drive the Viva Connections experience in Teams, referenced in greater detail at `https://docs.microsoft.com/en-us/viva/connections/sharepoint-app-bar`. The following figure shows us the app bar:

Figure 1.7 – SharePoint app bar

We've taken a look at the five core components of the modern experience in SharePoint Online. Understanding these elements will guide us through the rest of this book. However, to truly appreciate the modern world, we should spend a little time pondering the past and remember classic SharePoint.

Remembering the classic experience

Since we've defined key elements of the modern experience in SharePoint Online, we can now generally say that the classic experience is the one we find in SharePoint on-premises, especially between 2007 to 2016. For this chapter, let's collectively refer to the UI and features associated with it as SharePoint classic. Let's do so with the understanding that SharePoint 2016 has some of the modern features available in SharePoint Online and SharePoint 2019 has most of them.

Rather than reviewing all the templates, features, web parts, capabilities, and irritating quirks of the platform, let's review the top 10 components that either disappear entirely or become deprecated with a move to modern SharePoint Online:

- **Your own servers**: This might seem obvious, but it is highly impactful. Managing the infrastructure, including search, updates, failover, and so on, is no longer necessary. That means that we no longer have a **Central Administration** site, can't bump up the list view threshold, and don't have to worry about creating site collections that consume a new content database.

- **Full trust solutions**: Custom server-side code deployed via farm solutions or sandbox solutions is no longer supported in SharePoint Online, regardless of whether we have classic or modern sites.

- **Custom master pages and Cascading Style Sheets (CSS)**: Support for this resides only with classic sites. We'll look at this in more detail in *Chapter 3, Modern Options for Customizing SharePoint Online*, but suffice it to say that the concept of *branding SharePoint so that it's no longer recognizable* is a thing of the past.

- **Content Editor and Script Editor web parts**: The option to embed HTML, CSS, and JavaScript that impacts the entire containing page is no longer possible with SharePoint Online.

- **Classic site templates**: Specialized templates such as **Document Center**, **Record Center**, and **Search Center** are the exclusive domain of the classic world. Admins can create many of these in SharePoint Online, but the UI will be in classic mode.

- **Pages and publishing**: Wiki pages and web part pages are relics of the past and not possible with the modern UI. The Publishing Infrastructure features also remain in the classic world. We don't really need them with modern SharePoint with page sections, SharePoint News, and Power Automate to help us with approvals, but the features are still available in SharePoint Online with classic UI sites.

- **Search web parts**: The ability to add search boxes, results, refinement panels, and others is limited to classic pages, so we can't build our own search pages out of the box.

- **SharePoint Designer (SPD)**: The link to the tool is hidden from the UI of modern lists and libraries. By default, modern sites don't allow **Custom Script**, which means it wouldn't be supported anyway. Editing site pages in SPD doesn't really work, and workflows are now best created in Power Automate.

- **InfoPath**: This tool is the zombie of the SharePoint world. Officially dead (OK, deprecated, but the date of death has been announced) since 2014 but still walking around, consuming brains, and refusing to go away until support ends in July 2026. While InfoPath still works in SharePoint Online, we can't access it from the menu of modern pages, and the future lies in Power Apps.

- **Information management**: This one is a little different from the others in that it isn't part of a revamped UI, but is a function of SharePoint on-premises that goes away in favor of a richer alternative. Information management policies at the site collection, content type, or library level were built into SharePoint Server and provided retention, auditing, barcodes, and labels. This is now part of the Security & Compliance Center in M365 and provides a unified solution across content sources in the cloud, not just SharePoint.

While much of SharePoint remains intact, we can clearly see the changes Microsoft has made to frame the future for SharePoint. However, we can't speak of SharePoint Online in terms of exclusively classic or exclusively modern. The two worlds can live together.

SharePoint Online – a mixed skyline

SharePoint Online fully supports the modern UI, but it's not just moving to the cloud that provides its features. SharePoint Server 2016 and earlier do not support modern features or functionality. It is, however, possible to remain on-premises and have some of the modern UI elements as part of our SharePoint Server 2019 farm, which includes the following:

- Communication and team sites (though no connections to M365 Groups)
- Modern lists and libraries (though not Power Apps or Power Automate without being a hybrid)
- Site pages
- Modern web parts (though not the full list we see in SharePoint Online)
- Refreshed navigation and an M365-style app launcher (but no hub sites, app bar, or global navigation)

So, we can stay on-premises and still have some nice, new things – like watching an 8K TV from a 70s recliner. The flip side of that is we can move to SharePoint Online and still hold on to our classic treasures. The great C/M divide may be more of a blurry line than a solid one.

Maybe we can think of SharePoint sites as a mixed skyline. One of my favorite cities is Boston. Walking through the city, we can view a skyline filled with old stone churches and modern glass and steel structures. The change over almost 400 years is constantly on display, as seen in the following photo:

Figure 1.8 – Classic and modern buildings together in Boston

Microsoft tells us "*The 'classic' experience is not being deprecated; both 'classic' and 'modern' will coexist.*" That quote is taken from this guidance related to modernizing the UI: `https://docs.microsoft.com/en-us/sharepoint/dev/transform/modernize-userinterface`. Well... Latin is still spoken today, but there's little new and exciting coming to the world of Latin. Our goal should be to utilize modern SharePoint whenever possible, with eyes wide open to the fact that some classic pieces may still be needed.

In SharePoint Online, both classic sites and modern sites can coexist within the tenant. As a matter of fact, it's possible for modern sites to have classic pages and for classic sites to have modern pages. While we can't mix and match classic and modern web parts, the two worlds can live side by side. So, what really makes a modern site *modern*?

The short answer is the template we start with and the features we use. Modern team sites and communication sites create a place for us to work that has all the modern amenities. Greater detail about those amenities can be found here: `https://support.microsoft.com/en-us/office/sharepoint-classic-and-modern-experiences-5725c103-505d-4a6e-9350-300d3ec7d73f`. Starting with a modern template enables the following, thus enabling the modern experiences in SharePoint Online:

- The **site pages** feature is activated by default
- The home page is a modern page built from the **Site Page** content type composed of modern web parts
- The list and library UI is set to the new experience by default
- Modern search is utilized by default
- Modern site management is available from *the gear icon*, which allows us to see site performance, usage, modern theming, easier site permissions management, and site templates

Once that modern site is created, all new pages will be modern by default. We can, however, add classic pages as well. We can also leverage site collections and site features if we really need to retain some classic functionality. So, why do we want to keep a foot in both worlds?

In some remote instances, we must. One good example is the **Events** (or calendar) list. While there is an events web part to display them on a page, the list itself still renders in classic mode (old UI, ribbon, and all). I assume it's hard to get a table layout to be responsive.

We may also have invested a good deal of time and money in building out some highly customized pages that don't have a direct one-to-one translation to modern web parts. We may have a robust master page that will either take significant refactoring or a total redesign. There are certainly valid reasons to keep classic pages in place, though modernizing should be seen as an eventual requirement.

Let's look deeper into how the worlds can coexist from two perspectives. First, building a new modern site from scratch, and second, how to modernize a classic site.

The paths to modern

In our architecture planning, we need to understand the reasons why we need to migrate and the business value it can provide the organization. Considering the following points will help us ascertain who is involved in using modern sites, what they want to accomplish, and how they will use the features of modern SharePoint to collaborate:

- The number of content creators versus content consumers:

  - If a site is for a set of people who will be authoring and co-authoring content together, we should choose a team site

  - If a site is for a few content creators to communicate broadly across departments, to a region, or to the entire organization, we should choose a communication site

- The intended audience – invitation only or an open-door policy:

  - Team sites can either be private or public. Private sites can be extended to new members by having an owner invite them in. Public sites are open to everyone in the organization, but they have to walk in the door on their own. They're not automatically added.

  - Communication sites are more like classic sites in the way their permissions are managed, so the concept of owners, members, and visitors is essentially unchanged.

- The business needs to be met by creating a site:

  - What is the value and ROI of the site? Who will mainly benefit from the existence of the site? Who will be the content owner and caretaker of the site ensuring timely, accurate, and useful information?

These questions may help us to decide on a template. We can get these templates from the *look book* mentioned earlier, but applying a template after a site is made is now baked into the UI (`https://support.microsoft.com/en-us/office/apply-and-customize-sharepoint-site-templates-39382463-0e45-4d1b-be27-0e96aeec8398`). Microsoft provides a base set, automatically filtered to our site template, or custom templates unique to the organization.

The following figures show the UI for browsing for and adding a template to a site after it has already been created:

Figure 1.9 – Choosing a template when a site is created

Once we click **Browse templates**, we see the list of choices.

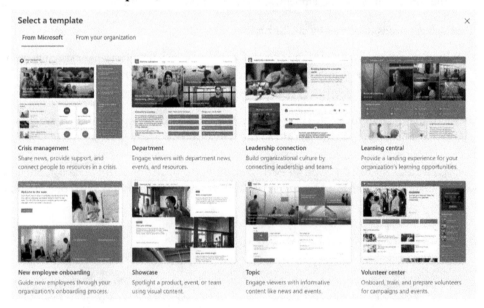

Figure 1.10 – Selecting a Microsoft site template

We have a couple more things to consider...

- The type(s) of interaction and collaboration expected:

  - While team sites are better suited for collaboration, the same capabilities exist between either site template type. As such, our choice is largely a matter of intent.

- The biggest exception to that is if you want to add real-time chat interaction via Microsoft Teams. In that case, we must choose the team site backed by an M365 group, so we can add that Teams workspace later on.

1. The role of this site in the larger organizational structure:

   - Is there a site that already exists that serves the same purpose? Does this site logically fit in with another that has similar people and purposes? For example, a Benefits site could likely accompany an HR site.

   - The real action we take from this information is to avoid duplication but also to plan our hubs. Sites that should logically work together can retain their individuality but join a hub to share resources. There will be more to come on that in *Chapter 7, Up with Hubs, Down with Subs – Planning Hub Sites*.

With responses to these points in mind, let's turn our attention to the options we have to move from classic to modern. We can start from scratch, migrate/convert a classic experience to modern, or quickly swap our root site to a modern alternative.

Modern from scratch

So, let's walk through this process of modern site creation end to end. As admins, we can create a new site in the SharePoint admin center. Any user, by default, can create a site from the SharePoint home page. To lock down that capability, we can modify the **Pages** setting in the admin center. The top checkbox in the following figure would need to be unchecked to disable page creation:

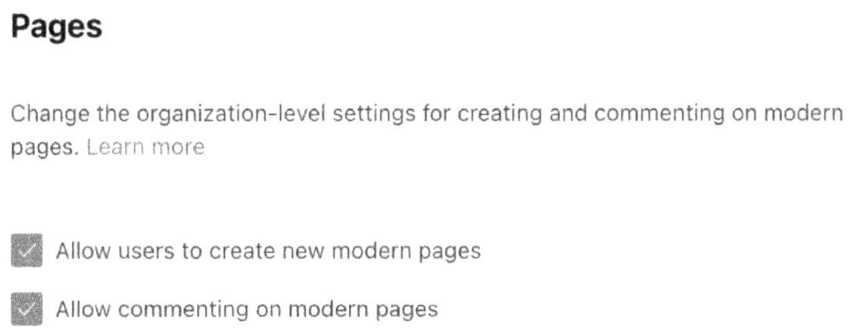

Pages

Change the organization-level settings for creating and commenting on modern pages. Learn more

☑ Allow users to create new modern pages

☑ Allow commenting on modern pages

Figure 1.11 – Org-level settings from the SharePoint Online admin center

Once we choose our template, the site will be provisioned within seconds. We then have the option to apply an additional template/site design that contains branding and content. That can be applied when we first visit the site or later in **Site Settings**. The template can be changed as many times as we need; this will impact the look and feel, but may also add list and library content. Any existing content won't be impacted or deleted. It will most certainly add a new home page and set it as the default.

Next, we can verify where our home page lives and where any additional pages will be created. Visiting site contents reveals the **site pages** library. It's here that we can also create classic pages as well as modern ones. In the following figure, we see that the modern **Site Page** and the classic **Web Part Page** and **Wiki Page** content types live together, allowing us to choose either when creating additional pages on our site:

Figure 1.12 – Menu when adding a new page from a site home page

The ideal experience, however, is to use the **Add a page** link from the *gear* settings icon or to use the **New** menu on the home page to create a new page. This option will allow us to choose a page template to scaffold our new creation. We can either use one of the Microsoft defaults, or we can save pages as templates, for use within the same site collection, back in the **site pages** library:

Figure 1.13 – Choosing a page template for a new site page

To explore further in the settings, we navigate to **Site Information** (seen in the following screenshot) to change the logo, title, description, and privacy settings for team sites that may need to go from open to closed or vice versa. The classic site settings page with all the usual configuration options still exists for all site types and is accessible by clicking the link shown at the bottom of the screenshot (**View all site settings**):

Figure 1.14 – Site settings for a modern site

Site usage takes us into an analytics page, which provides a view into the amount of traffic, popularity of the content, types of devices, and how long people spend on our sites. This data can show up to the last 90 days. Site performance is a relatively new addition that provides some instruction and a download link for the **Page Diagnostics for SharePoint tool**, which is an add-on for Microsoft Edge browsers. This tool provides deeper technical debugging and performance information. It is available at `https://aka.ms/perftool`.

Modern from classic

For sites that contain classic pages, we can choose to leave them in place for now or plan to modernize them. These pages will be based on classic Wiki or Web Part content types and contain classic web parts. For a time, the root site collection in our tenants was a classic site, though that default would later change. As such, many organizations built classic sites as the root of their intranets in SharePoint Online. Many organizations have also migrated from SharePoint on-premises to the cloud, leading to sites remaining in classic mode.

If we do plan to modernize classic pages, we can choose between three options:

- Recreate the page from scratch

- Rely on automatic home page modernization

- Use the Page Transformation UI detailed in this reference from Microsoft: `https://docs.microsoft.com/en-us/sharepoint/dev/transform/modernize-userinterface-site-pages-model`

Automatic modernization only works within a narrow set of parameters defined here: `https://docs.microsoft.com/en-us/sharepoint/disable-auto-modernization-classic-home-pages`. If your home page is classic, it can be modernized automatically for classic team sites, but no other template. The page has to be called `home.aspx` and must only contain default web parts with no other content or customization. I see this being most useful in migration scenarios. We can also leverage PowerShell to turn classic team sites into communication sites, complete with full-width pages: `https://docs.microsoft.com/en-us/powershell/module/sharepoint-online/Enable-SPOCommSite`.

Recreating pages from scratch may seem daunting but it can be a very positive approach. Doing so gives us a great excuse to excise technical debt (to stop holding onto things created in the past that no longer work well or fit the need). It can also give us a chance to revisit the purpose and desired content of the page with business stakeholders. Their needs may have changed or the stakeholders themselves may have changed since the site and pages were first created, and starting from scratch gives us a moment to start fresh with updated requirements.

The **Page Transformation** UI is not an out-of-the-box product from Microsoft already available in your tenant. It is instead a part of the SharePoint PnP Modernization solution built on the PnP Framework: `https://github.com/pnp/pnpframework`. The solution essentially takes web part mapping in an XML format and programmatically translates web parts from classic to modern using PowerShell or .NET. The process is that a transformation model file is read, a classic page is analyzed, selectors are made to find impacted page elements, and the mapping is applied, resulting in a new, saved page (`https://docs.microsoft.com/en-us/sharepoint/dev/transform/modernize-userinterface-site-pages-model`).

This may be a solution for an environment where there is a large volume of sites and pages. The effort is done more up-front by developers to reduce the burden on content owners or creators who may not have the time to do the work of building from scratch. In addition to page mapping, there are also options to map users, metadata terms, page layouts to sections, and so on. Should you choose to go down this route, Microsoft provides a library of scripts to get you started quickly: `https://docs.microsoft.com/en-us/sharepoint/dev/transform/modernize-sample-scripts`.

Swapping to modern

As noted previously, your home site may have been created using a classic site template. This is likely true if your tenant was created prior to April 2019. While it is easy enough to modernize it by adding the elements we've talked about so far, we may have spent time building a modern site elsewhere. A pretty common scenario I've seen is for a site, essentially serving as an intranet site, to have been created separately from the root in the `/sites` path. There has often been some type of JavaScript or another redirect to load that site when someone visits the root URL.

If that is the case in your tenant, you can run a site swap, either through PowerShell or in the SharePoint admin center, to modernize the home site by replacing the root with a site in a different location. The old site is migrated to a new location, which can serve as an archive until we're sure everything is good to go. This is also a great way to refactor your home site in chunks over time while still keeping a single big-bang reveal. The following figure shows us a screenshot from the SharePoint admin page allowing us to swap a new, modern site into a classic root site:

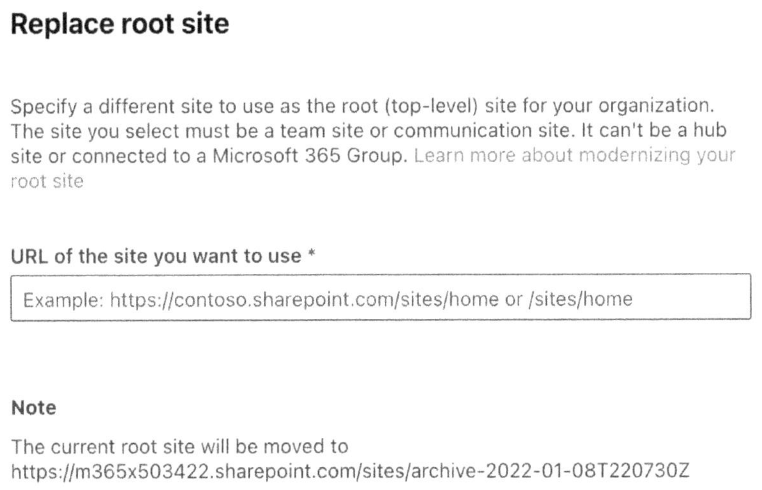

Replace root site

Specify a different site to use as the root (top-level) site for your organization. The site you select must be a team site or communication site. It can't be a hub site or connected to a Microsoft 365 Group. Learn more about modernizing your root site

URL of the site you want to use *

Example: https://contoso.sharepoint.com/sites/home or /sites/home

Note

The current root site will be moved to https://m365x503422.sharepoint.com/sites/archive-2022-01-08T220730Z

Figure 1.15 – A screenshot of site swap in the SharePoint admin center

Essentially, we are backing up the current root site (at the root URL for the SharePoint tenant) and then migrating a different site into that URL. To make that happen, our new soon-to-be-root site cannot be connected to an M365 group, can't already be a hub site (or must have associated sites removed), and shouldn't contain subsites. On the active sites page of the SharePoint admin center, we would need to select our root site and provide the URL for its proposed replacement, as in the preceding screenshot.

We've seen three possible options for moving from classic to modern. These options make sense if we plan to wholly replace classic with modern. That is a solid long-term goal, but we may want or need to have our on-premises farm connect to services running in the cloud.

Hybrid workloads

So, we have classic and modern, on-premises and online, Marvel and DC, and pizza and burgers. Sometimes it's hard to choose just one. For those of us who still have an on-premises SharePoint farm but want to modernize, maybe we don't have to. SharePoint Server 2013, 2016, and 2019 all provide some support for hybrid workloads, which allows us to leverage the best of both worlds. Remember, only SharePoint 2019 gives us the option to implement the modern UI, however.

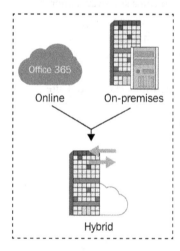

Figure 1.16 – Hybrid combines on-premises and online

As part of our planning to move to SharePoint Online, we may find that certain compliance or security concerns or the need for connections to other systems dictate that we at least keep some kind of on-premises presence. Microsoft has some great guidance here: https://docs.microsoft.com/en-us/sharepoint/hybrid/plan-sharepoint-server-hybrid. A few key ideas to consider when planning for hybrid follow. These represent the most widely used and perhaps the most valuable hybrid workloads.

Hybrid search

If your content is going to live in two worlds, having a unified way to find what you need will become crucial. A hybrid search is available in SharePoint 2013–2019. Cloud hybrid search and hybrid federated search are two options for implementing this. The former indexes all your contents in the cloud. The latter creates two indexes that can feed search results from both places in response to a single search query.

So, both solutions provide results from both places. The question is whether we want to have two different search results with their own rankings, refiners, and others, or one single combined set. We may also need to make sure we plan for content that is highly sensitive that doesn't need to be indexed in the cloud. For that, we would choose the federated route.

Hybrid OneDrive

This feature allows you to leverage OneDrive for Business for personal work-related content in the cloud while keeping your SharePoint site on-premises. This may be a great option for organizations that are slowly moving away from their farms to M365 but are not able to move everything at once.

Our users could take advantage of 1 TB of cloud storage per user, accessible from any device that can be shared externally, while we plan for SharePoint migrations over time. The configuration here would redirect the on-premises link to a user's My Site to the cloud instead.

Hybrid app launcher

The last hybrid workload we should note at this point helps us to have a consistent user experience when navigating. The app launcher, or waffle, in M365 is a grid in the upper left that opens a list of apps and services we have access to. It is really the global navigation across the M365 toolbox.

Enabling this feature on-premises allows custom tiles you've pinned to exist side by side with the same list of apps your users would see if they were browsing SharePoint Online. This feature is connected to **Hybrid Sites** in SharePoint 2016, which includes site following and profiles as well.

Additional features

There are a few additional features we need to review to understand the possibilities in modern SharePoint. These features provide a backbone for communication and corporate identity within and across modern SharePoint sites:

- News pages
- Dynamic web parts
- Headers and footers
- SharePoint domain name

News pages are just **site pages** with the **Promoted State** metadata column equal to 1 or 2 (1 is unpublished news, 2 is published, and 0 is a standard site page) and an additional metadata column of **First Published Date**, which identifies their purpose to SharePoint. They can be built in the same way as other modern pages, though their content is often simpler and more focused on text than images. The real power is on the receiving end.

The **News** web part can gather news stories from across sites in SharePoint Online, including those within a hub, or sites that we manually include. This is great for an intranet home page because content creators can generate news from their own sites (where they have edit access) and have those automatically roll up to a central location.

In addition to the **News** web part, we can also see that same news published on the SharePoint home page where we can see news from all the sites we have access to. The SharePoint mobile app rolls up the news in a similar way and can provide push notifications to let users know something new is available. News can also be surfaced in Microsoft Teams tabs or channel chats via a connector.

Given all the places news can surface, it certainly beats sending all-company emails and would be available historically for new employees to see, even if they joined after it was created. By the way, we can still go into a news item or the news digest and send those emails out (with links to the story) if we can't break that habit.

Other web parts that keep our sites fresh and compelling with dynamically updated content include the following:

- Weather and Twitter web parts that connect to external data
- The **Recent Documents**, **Sites**, and **My Feed** web parts provide personalized content unique to each user viewing the site
- The **Events** web part can roll up SharePoint calendar events from across sites in a similar fashion as the **News** web part

Microsoft has compiled a single web resource to help us understand the entire set of modern web parts: `https://support.microsoft.com/en-us/office/using-web-parts-on-sharepoint-pages-336e8e92-3e2d-4298-ae01-d404bbe751e0`.

Modern headers and footers give us the option to create consistent site experiences. In addition to the logo and title we've always had on SharePoint sites, we can also choose to have a color or an image across the top of our pages to put our corporate branding front and center. If we prefer a more minimalist approach, we have that option too, plus the user can choose to expand or collapse the header on the fly when they view our site. In *Chapter 7*, we'll look at hub sites as a way to keep the consistency of the color theme across a set of connected sites.

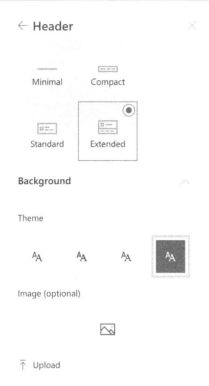

Figure 1.17 – Changing the look on a modern site

We can add a site footer to communication site pages created from the **Site Page** template (which would exclude list/library pages). This provides a band along the bottom of the page with colors matching our branding, a small logo, and some text-based hyperlinks. This would have taken some customization to accomplish in classic SharePoint.

The final feature of note doesn't impact the UI but can have a significant role in corporate identity. When a tenant is created, the domain portion of the URL for SharePoint is the same as the tenant name. Until recently, we've been stuck with that unless we go through a painful tenant-to-tenant migration. Now, we can change the tenant portion of the name for SharePoint to better reflect our needs. For example, if we created our tenant as `coolstuff2021.sharepoint.com`, we can change it to `CoolThings.sharepoint.com`.

This may come in handy if the organization has gone through a rebranding, a merger with another company, or an acquired company needs to take on the new parent's organization branding. While our tenant name, email addresses, and so on do not change, we can update the SharePoint and OneDrive URLs once every 6 months. If the need arises, the most up-to-date guidance from Microsoft may be found here:

```
https://docs.microsoft.com/en-us/sharepoint/change-your-sharepoint-domain-name
```

Now that we've explored some key differences between the classic world of SharePoint on-premises and the modern world of SharePoint Online, let's take a look at some planning considerations for configuring and governing SharePoint in the Microsoft cloud.

IT governance

Ask 10 people what *IT governance* means and you're going to get 11 answers. Sometimes we picture a single, large, and dusty document that contains all the rules we should follow. Many erroneously think of security when governance is mentioned. Most simply glaze over or quickly change the subject.

From my perspective, governance is the gatekeeper of good adoption. One small component of any good governance plan is making sure that the service configurations are aligned with best practices and business expectations.

While SharePoint will be configured with default settings and users will automatically have a SharePoint and OneDrive license enabled as part of their M365 license entitlements, we may want to verify some of those settings to ensure it is adopted in the way we desire.

External sharing

Should our users be able to share documents stored in SharePoint or OneDrive for Business sites with those outside the organization? In the SharePoint admin center, under **Policies | Sharing**, we can set up the guardrails with the option to alter the setting for individual sites if the need arises. We can also decide to only let users within specific Azure AD security groups share content externally (maybe vendor management, sales staff, etc.).

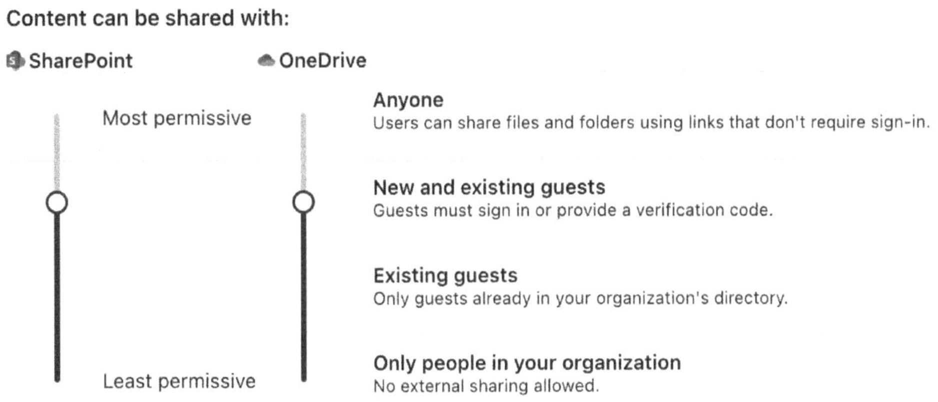

Content can be shared with:

SharePoint   OneDrive

Most permissive

Anyone
Users can share files and folders using links that don't require sign-in.

New and existing guests
Guests must sign in or provide a verification code.

Existing guests
Only guests already in your organization's directory.

Only people in your organization
No external sharing allowed.

Least permissive

Figure 1.18 – Sharing permission settings for SharePoint and OneDrive

As part of our planning, we need to decide whether our users will be able to add new guests along the way by simply entering their email, or whether guests (anyone outside the organization) should be added in advance to Azure AD by an IT admin. That is the primary difference between the second and third options in the list in the preceding figure. The last option in the list limits sharing to only licensed users in the tenant. The first option should rarely, if ever, be used, in my opinion.

Setting the default experience for sharing links can go a long way toward driving the right adoption habits as well. Though site owners can alter the defaults for their site and the options can be changed each time something is shared, changing the defaults here is a definite recommendation. By default, we have these options selected:

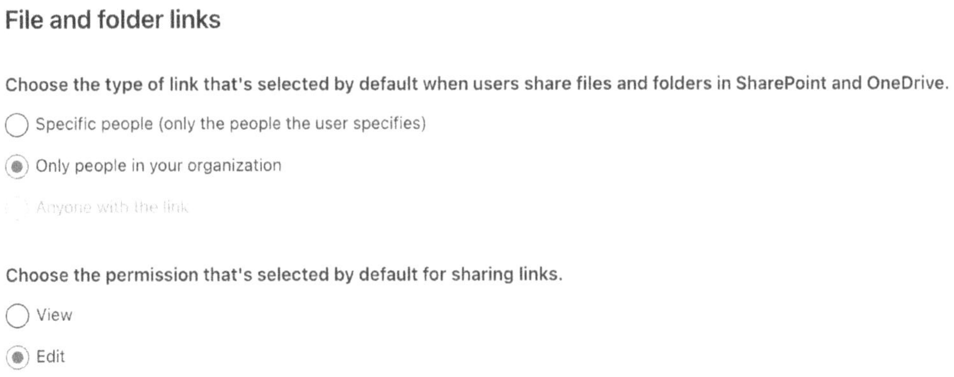

File and folder links

Choose the type of link that's selected by default when users share files and folders in SharePoint and OneDrive.

◯ Specific people (only the people the user specifies)

⦿ Only people in your organization

◯ Anyone with the link

Choose the permission that's selected by default for sharing links.

◯ View

⦿ Edit

Figure 1.19 – Options in the sharing dialog for SharePoint and OneDrive

This is pretty much like sharing something with everyone at once. If links are forwarded, they'll work for anyone in your organization that opens the link. Having **Edit** selected means that anyone who opens a document will be able to make changes to it right away. My experience has been that we are better off defining specific people with whom we want to share, and only allowing them to view by default, making co-authoring the exception when sharing. These may be better default values to start with (with a deeper dive found at `https://docs.microsoft.com/en-us/sharepoint/turn-external-sharing-on-or-off`):

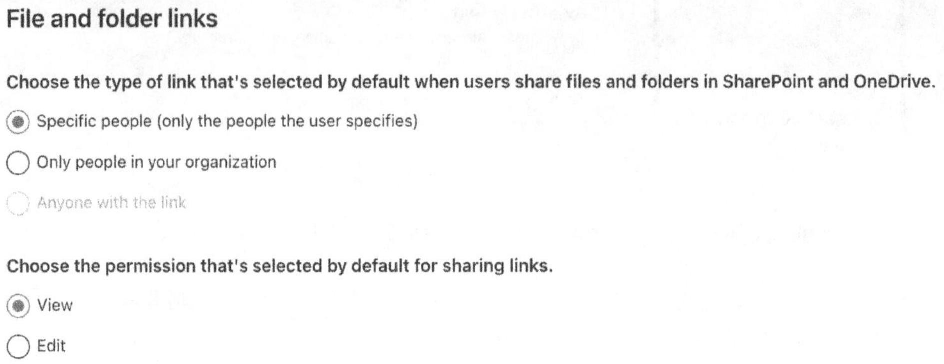

File and folder links

Choose the type of link that's selected by default when users share files and folders in SharePoint and OneDrive.

- ● Specific people (only the people the user specifies)
- ○ Only people in your organization
- ○ Anyone with the link

Choose the permission that's selected by default for sharing links.

- ● View
- ○ Edit

Figure 1.20 – Same options with specific people and view options chosen

Settings

A handful of settings can make a significant impact on governance and adoption. In the admin center, when we click on **Settings**, we get the following view:

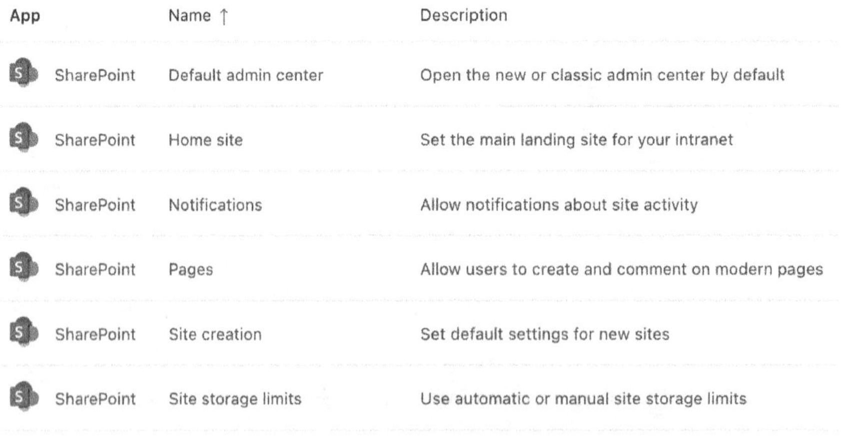

App		Name ↑	Description
S	SharePoint	Default admin center	Open the new or classic admin center by default
S	SharePoint	Home site	Set the main landing site for your intranet
S	SharePoint	Notifications	Allow notifications about site activity
S	SharePoint	Pages	Allow users to create and comment on modern pages
S	SharePoint	Site creation	Set default settings for new sites
S	SharePoint	Site storage limits	Use automatic or manual site storage limits

Figure 1.21 – Settings listed in the SharePoint admin center

We've already discussed the value of the **Site creation** setting, where we can limit whether users can create new modern sites (what we would have called *site collections* in classic SharePoint). We can globally limit whether users should be able to create new pages. When disabled, anyone with permissions could still create **News** pages, but we'll probably be better off leaving this on and allowing it to be managed by site owners. Controlling permissions to the **site pages** library is likely our best bet.

There may be some value in turning off comments globally, however. These comments are unmoderated and only scratch the surface of providing the kind of engaging experience that Microsoft Teams chats or Yammer discussions could provide.

The other setting we should be mindful to plan for is the home site. The setting just takes a URL (most likely the root intranet site), but the significance behind it impacts several future capabilities. Setting the home site is going to be crucial for enabling Viva Connections later on. It is also necessary for the app bar to work and automatically gives us a site that acts as an official organizational news source (`https://docs.microsoft.com/en-us/sharepoint/organization-news-site`).

Summary and planning document

In this chapter, we've learned about the building blocks of a modern SharePoint site. The five core blocks are sites, lists and libraries, pages, web parts, and navigation. Classic pages in SharePoint use a classic set of components, while modern pages use a different set. This new set provides a responsive design that is much more user-friendly for editing and maintaining by a business user, not by IT.

We explored the classic world of SharePoint, which can still coexist with modern in the cloud. We discussed potential paths for moving from classic to modern. For environments where an on-premises presence can't be entirely removed, we looked at the possibility of hybrid workloads. We concluded with a note about IT governance with an eye toward sharing capabilities that those new to the cloud will find exciting but possibly confusing.

At the close of each chapter, I want to distill the information we've reviewed while also giving you a template to create your own SharePoint architect's planning and governance guide. This is really just a recap of the pieces we should be accounting for when moving to SharePoint Online, either from a classic past or as the first foray into a Microsoft cloud future.

Configuration and governance

The following is what we have learned about configuration and governance:

- **External sharing**:
  - Which business units or users require sharing outside the organization?
  - What should the global default be?

- **Site creators**:

  - Should everyone be able to create a site when they want it?

  - Should IT create all sites in the admin center?

  - How do users make a request if the self-service option is removed?

- **Page comments**:

  - Should comments be allowed?

  - Is there another tool available to capture and moderate comments and discussions?

Site planning

The following are the notes to remember about site planning:

- **Home site**:

  - Has a home site been created and set in the admin center?

  - Are you using the root site as the home site?

  - Do you have a classic site or modern site at the root URL?

  - Have you built a site elsewhere that should be swapped to the root URL to serve as the home site?

  - Does the SharePoint domain in the URL reflect the organizational identity or does it need to be changed?

- **Departmental sites**:

  - Does each department/location/organizational unit have a SharePoint site?

  - Does a site need to be available to just users inside the department or is it for org-wide consumption?

  - Does the home page of the site need to be modernized?

- **Site owners**:

  - Who should be the owners of new sites?

  - Will site owners and content creators be the same?

- **Pages**:

  - Should both classic and modern pages be allowed on new sites?

  - Are there pages in classic mode that need to be modernized?

- Are there pages that must remain in classic mode due to the presence of and need for customizations?

- Are important pages reflected in the site's navigation?

- Who is responsible for creating official SharePoint news?

Architecture

The following are the notes to remember about architecture:

- **Hybrid**:

  - Are there still on-premises SharePoint servers?

  - Do we need to keep any workloads or data on-premises?

  - Do we have a production M365 tenant with other workloads configured?

  - Can any of those workloads be integrated with our on-premises farm?

  - Do we need to configure a hybrid search? If so, where should the index reside – on-premises or cloud?

- **Personal sites**:

  - Do all users have a OneDrive site that has been provisioned?

  - Are all users aware of their OneDrive for Business storage?

In the next chapter, we will explore the process and tooling that will help us plan to migrate our content from on-premises to the cloud. This will give us a complete picture of the steps needed to get into the modern SharePoint world, which will set up the remainder of this book and allow us to focus on deeper details to build our planning guide for SharePoint and Office 365.

Further reading

- Learn more about Sharepoint here: `https://docs.microsoft.com/en-us/viva/connections/sharepoint-app-bar`

- Details about classic and modern amenities are available here: `https://support.microsoft.com/en-us/office/sharepoint-classic-and-modern-experiences-5725c103-505d-4a6e-9350-300d3ec7d73f`

- More details about SharePoint site templates are available here: `https://support.microsoft.com/en-us/office/apply-and-customize-sharepoint-site-templates-39382463-0e45-4d1b-be27-0e96aeec8398`

2

Making the Move – Migration Options and Considerations

In order to bridge the **classic/modern (C/M)** divide, we must be prepared to leave the past behind and boldly embrace the future. In *Chapter 1*, we looked at all the new capabilities in modern SharePoint sites and pages. Our plan should be to always use modern for anything new. However, we may need to pack up our old sites and drag them kicking and screaming across the divide. This chapter is the last that will focus on classic SharePoint. In order to leave classic behind, we need to get ready for a *SharePoint migration*.

We will go through the following in this chapter:

- Moving day – getting ready to migrate
- Picking the right movers – migration tools
- Potential problem spots
- Helper tools
- Summary and planning document

Moving day – getting ready to migrate

Moving from an old house to a new one can be equally exciting and exhausting. There are so many details to manage to make the move a success. Things have to be packaged with care so that nothing gets broken; movers have to be hired, schedules have to be coordinated, and items have to be unpacked over time and put in just the right spot in the new place. A successful move must follow a series of steps in the right order, which means we need to start with a good plan.

SharePoint migrations are very much like that move to a new place. In my experience, these migrations have usually been one of three varieties:

- On-premises to cloud
- Tenant to tenant
- File server to cloud

The rules may differ in their specifics, but the process is generally the same. We'll review how each one differs in a bit, but for now, let's walk through what is always needed with a migration:

- **Inventory**: What items need to be moved? Do certain items need special handling? In a home move, maybe that's a piano or a safe. Did we remember all the stuff in the garage or the storage shed? In SharePoint, this may mean sites, files, customizations, metadata, and so on, but having a clear list of what we're moving is a crucial first step.

- **Cleanup**: The importance of this step simply cannot be overstated. It tends to be one of the most time-consuming, and so one that is often skipped. It's always amazed me how much clutter we gather over the years that only gets recognized as such when it's time to move. Many organizations have a *keep everything forever* policy because it is easier than having to evaluate and decide what to discard. The more we toss out, the more efficient the move.

- **Mapping**: Maybe the new place we're moving to is a different size than our old one. Whether it's bigger or smaller, a change in location causes us to map out where our stuff is going to go in that new space. We've done our cleanup but items that serve a purpose or have sentimental value need to come along, and we need to figure out the best spot for them. With SharePoint migrations, we need to map sites, lists/libraries, folders, users, permissions, and others so that our migration target is configured optimally, not just lifting and shifting the same old stuff.

- **Relocation**: This is the actual process of moving your stuff from one spot to the other. Depending on the complexity and effort expected, we may try the move ourselves or hire someone to do it for us. Maybe we move everything in one big truck or take several small trips over days or weeks. SharePoint migrations can be done all at once, but often, more time is needed. Even if the amount of data is manageable, we may want to limit the impact on the user experience. A phased migration often works best.

- **Validation**: We need to make sure everything from the inventory made it to the new place and that it is all still in one piece. With SharePoint data, the IT engineer moving the contents can verify by doing a spot check and checking logs, but it is really the content owner who needs to verify that all their information is there, that their pages still look right, and workflows or other features still function correctly.

- **Communication:** When we move to a new place, there are a lot of people who need to know the new address or directions. The move itself needs good coordination, with the movers, utilities, and deliveries all needing to be informed. This part of the process isn't really a component unto itself but rather a companion for all the others. Migrations often fail more from a lack of communication than from technical errors.

Now that our process is defined so we can repeat it, let's turn our attention toward selecting the right tool(s) to make the move as simple and successful as possible.

Picking the right movers – migration tools

When moving house, we may need some help with packaging, having the right truck, and having movers to help us do the heavy lifting. With SharePoint migrations, selecting the right tool is equally important. Whether it's a self-service move or we have help from consultants, a migration tool is crucial. That tool should assist us with all five aspects of a successful move we noted in the previous section.

I've been involved in numerous migrations over the years. To handle those migrations, there have been a couple of third-party tools I've relied on. The most common tools I've used in recent years have been products from ShareGate, BitTitan, AvePoint, and Cloudiway, depending on what needs to be moved (especially if the move involved Microsoft Teams from one tenant to another). However, I want to focus on three toolsets available to us as part of our **Microsoft 365** (**M365**) or **Office 365** (**O365**) subscriptions – the tools provided by Microsoft. The following infographic from Microsoft summarizes the paths that we have in front of us, all determined by the source of data to be migrated:

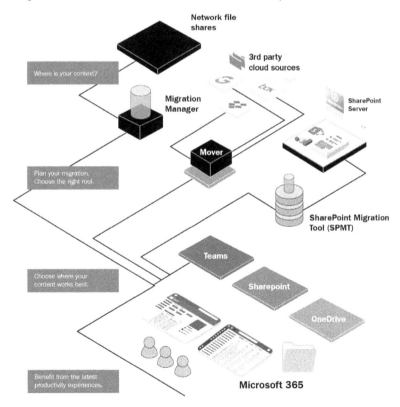

Figure 2.1 – A conceptual map for migration sources and tools

A variety of tools is available to give us the best option for a variety of sources. Let's take a look at each tool individually to understand what it does well and where it won't help.

The SharePoint Migration Tool (SPMT)

SPMT may be downloaded from the SharePoint Online admin center. This is the right tool for migrating SharePoint Server 2010-2016 and SharePoint Foundation to SharePoint Online. SPMT can also be used to migrate MySites (a site collection for personal data) to OneDrive for Business.

The full list of supported features can be found here: `https://docs.microsoft.com/en-us/sharepointmigration/what-is-supported-spmt`.

SPMT is a locally downloaded application, available only for PCs, which allows us to define connections to the on-premises farm with at least read access to the content, as well as the destination tenant as a SharePoint or Global Administrator. The prerequisites for the tool are as follows:

CPU: 64-bit 1.4-GHz 2-core processor or better

RAM: 8 GB

Local storage: hard disk, 150 GB free space

Network card: high-speed internet connection

Operating system: Windows Server 2012 R2 or Windows 10 client or later and .NET Framework 4.6.2 or later

The recommended requirements defined by Microsoft for best performance are as follows:

CPU: 64-bit quad-core processor or better

RAM: 16 GB

Local storage: solid-state disk, 150 GB free space

Network card: 1 Gbps

Operating system: Windows Server 2012 R2 or Windows 10 client or later and .NET Framework 4.6.2 or later

The tool gives us the option to keep file versions, filter migration files by date or file type, migrate OneNote notebooks, and migrate site collections, sites, pages, lists/libraries, content types, and managed metadata terms. Classic pages will look the same in the SharePoint Online destination as they did in the source, so a manual refactoring of those pages will be necessary.

How does SPMT help us to address the five aspects of a successful move?

Inventory

There is a companion tool called the **SharePoint Migration Assessment Tool** (**SMAT**), which can initiate a scan of a SharePoint farm that runs over a few days and generates output spreadsheets, providing a thoroughly detailed inventory of every aspect of your farm. This includes data as well as apps, customizations, workflows, and potential pain points such as large sites and long OneDrive URLs. We're going to find individual scanner reports revealing details about components that possibly won't migrate at all or may prove difficult to migrate. We'll get a listing of the following key elements, among others:

Apps

User alerts (which will have to be recreated)

Business Continuity Services (**BCS**) connections to external data (which will likely be replaced with an M365 connector)

Checked-out files (which will have to be checked in to fully migrate)

Customized pages (which won't migrate to modern)

Email-enabled lists (a function no longer supported in the cloud)

External lists (which use BCS connections to expose external data)

InfoPath (which will migrate but is deprecated)

Information Rights Management (**IRM**) enabled lists (which may migrate but should be updated to use M365 IRM)

Large files, lists, and sites (the larger the source, the longer the migration)

Large list views (over 5,000 items will likely fail in the cloud)

Master and publishing pages (which only work for classic sites)

Sandbox solutions (which won't migrate but may need to be rebuilt)

Workflow associations (which tell us where workflows are being used)

The description of all the files that will result from the scan and the information that is captured can be found here: `https://docs.microsoft.com/en-us/sharepointmigration/ sharepoint-migration-assessment-toolscan-reports-roadmap`. Log files containing information, warnings, and errors are also produced.

Cleanup

Ultimately, the decision about what data to keep and what to delete boils down to the desires of the content owners. The tool offers some assistance for those wanting to reduce the amount of data that gets migrated over. These tools assist in that cleanup effort by helping you to better identify content that is older, obsolete, or redundant. Content should be deleted at the source rather than migrated and then cleaned up later. Choosing to filter files by created or modified dates may help us to only bring over the most recent content. This also provides a way to limit the size of the migration, which in turn will allow it to begin and end more quickly.

Mapping

In addition to the full SMAT scan, there is also a mode for identity mapping that reports on all users and groups in place on-premises and creates an initial mapping of those identities to ones in Azure AD.

The source site and list/library may be specified for each migration task along with the target site and list/library. SPMT also offers a bulk upload process whereby you can use a CSV or JSON file to identify each migration task as a line or a row. By populating the columns in the spreadsheet, we can map a portion of a migration or the entire migration and use the spreadsheet as a task runbook. The columns we can specify are as follows (and detailed here: `https://docs.microsoft.com/en-us/sharepointmigration/how-to-format-your-csv-file-for-data-content-migration`). If only the source site URL and the target web URL are defined, all data in that site will be migrated:

- Source site URL
- Source document library
- Source subfolder (a folder inside the document library)
- Target web URL in the tenant
- Target document library
- Target subfolder

We can also populate `RegisterAsHubSite` and `AssociateWithHubURL` to begin the mapping of our old structure to a new modern structure. We'll discuss hub sites in *Chapter 7, Up with Hubs, Down with Subs – Planning Hub Sites*, in more detail.

Relocation

This one's pretty easy. SPMT can do the job for both SharePoint and OneDrive/MySites data. We can either do this one task at a time (using the bulk upload tool just discussed) or via PowerShell.

One key concept to keep in mind with any migration, however, is that whether we use scripts or the tool, we are leveraging remote APIs that are sharing the same bandwidth and throughput as all of the other requests for data from our cloud-based platform. Microsoft throttles background apps, such as migration tools, during weekday daytime hours. Evening and weekends then become the optimal time to run your migrations and have them run faster. Additional guidance for a better-performing migration can be found here: `https://docs.microsoft.com/en-us/sharepointmigration/sharepoint-online-and-onedrive-migration-speed`. We can also take advantage of the performance report to predict how fast our migrations will perform.

Validation

After the migration is over, we have access to several summary and task item reports. These include a failure summary and an item failure report. These reports allow us to better understand errors that occurred during a particular migration task. We can use these reports to fix those problems or make changes to content locking, filename issues, and path length issues. A review by the content owners following the migration is still invaluable, but we should be able to keep the issues they see to a minimum.

Migration Manager

This is the tool of choice for migration from network file shares to SharePoint libraries but also provides content migrations from Box, Google Workspace, Dropbox, and Egnyte. Migration Manager now includes what used to exist in the `Mover.io` web app. We can access this web-based tool through the SharePoint admin center.

If the focus is on network files, we'll need to provide the path to the file shares and an account to read those files, as well as a global or SharePoint admin account on the target tenant. Full and incremental migrations are possible.

For cloud sources, we need to make some connections with the appropriate logins. The connectors generally result in an app being authorized in the source environment with the assistance of an API key. In my experience, this might take some persistence. The documentation is thorough, but I've found that making sure you're connecting as *the right kind of admin* and having the source accept the app registration is trickier than it seems.

Inventory

A scan is performed automatically once a source file server location is added. That scan produces reports that can be viewed to validate your inventory and to identify any files that will be problematic when running the actual migration. Visualizations within the tool will show the overall size, the number of files, and warnings.

Cleanup

We can identify a subset of files to include in our migration tasks. From my experience, cleanup on file shares can be the most difficult. This is partially because that file share may contain decades of files. The usual file retention on shared drives is infinity plus forever. As a result of the *keep everything forever* mindset, it is very difficult to have a meaningful purge of files before a migration begins. It is always crucial to only bring over what truly matters.

Mapping

The mapping story here is mostly the same as SPMT. The user mapping and bulk upload tools are essentially the same. The added complexity here is that not all files are neatly separated into user versus shared access. With MySites, we are confident those will go into SharePoint Online. The mapping is clear. SharePoint on-premises sites usually map one for one or into easily defined replacement sites. With file share migrations, we may have something like a *U Drive* for user data, where each person has a dedicated folder. There may be an *S Drive* or something similar, with shared data broken down by business unit. Where that delineation is not clear, however, will require a more extended mapping exercise before we set up migration tasks.

Relocation

Multiple agents on multiple machines can certainly help increase the efficiency of migrations. The infrastructure of your network, the size and number of files, and throttling on the cloud target side can all play a part in slowing down that process, however. The **Migration Performance** section of the Migration Manager tool can provide a historical view of migration performance across our timeline. The following figure shows a migration performance data visualization (noted in greater detail in the source at `https://docs.microsoft.com/en-us/sharepointmigration/mm-performance-dashboard`):

Migration performance

Migration performance can be impacted by network infrastructure, file size, migration time, and throttling. Understanding these will help you plan and maximize the efficiency of your migration. Learn more

Performance history

View: **All** ∨   Type: **Files and folders** ∨   Date: **04/12/2021-04/15/2021** ∨   ✕

Recommended improvements

Everything looks good! However, if your performance changes, learn more about the factors that impact migration performance.

Figure 2.2 – A migration performance graph from Migration Manager

The performance graph helps us to see a representation of the performance of the overall migration refinable by content type and date range.

Validation

We have the same reports available with Migration Manager as we have with SPMT. Multiple reports will be produced as a ZIP file that contains information about what worked, how many items were moved, and where errors occurred. This link provides details on those reports: `https://docs.microsoft.com/en-us/sharepointmigration/mm-reports`.

Helper tools

Every good mover brings some helper tools along to get the job done more efficiently. The right dolly, some furniture pads, or cellophane to wrap things tight makes a mover's life simpler. Let's mention two tools here that make moving data into SharePoint Online and modernizing it a little easier as well.

The SharePoint modernization scanner

A common path for customers I've served has been to lift and shift their SharePoint on-premises farms to the cloud using a migration tool. The only real modernization that happens is on lists and libraries, where the change is as easy as the flip of a switch to display them in the modern UI. This means that links to InfoPath forms or SharePoint Designer workflows that are still hanging around can no longer be seen in the UI, though they may still function behind the scenes.

When that lift and shift migration is complete, there may be many sites that are still in classic mode. If the organization has adopted Teams, they have (perhaps unknowingly) created many modern SharePoint Team sites behind the scenes. Their skyline may be mixed more than they realize. Enter the SharePoint Modernization scanner, a tool that can be run against a SharePoint Online source to help you identify where modernization is lacking. The SMAT tool we noted earlier can provide an analysis of a source before content is migrated. The Modernization scanner is a supplement to that and is designed to be executed once a migration is complete.

Microsoft tells us at this site, `https://docs.microsoft.com/en-us/sharepoint/dev/transform/modernize-scanner`, that the following elements are identified in scanner reports:

- Optimizing the usage of modern lists and libraries
- Connecting these sites to a Microsoft 365 group
- Modernizing the Wiki and web part pages by creating modern site pages
- Rebuilding classic publishing portals as modern publishing portals
- Understanding where classic workflow is used
- Understanding where InfoPath is being used
- Understanding the usage of the classic blog page

We can use the scanner to scan a full tenant or a set of site collections we choose. For bulk scanning, we can also supply a listing of site URLs in a CSV file. The resulting reports are in an Excel spreadsheet with some visual dashboards already designed to help us see the results at a summary level. The reports can give us insight into sites that are ready to **groupify** (which means that we can connect them to an M365 Group, which we'll cover more in *Chapter 4*), which sites have uncustomized home pages, where publishing features are used, workflow and InfoPath usage, and which lists have customized forms (which may have been in InfoPath or SharePoint Designer). An example of one of the generated dashboards is shown here:

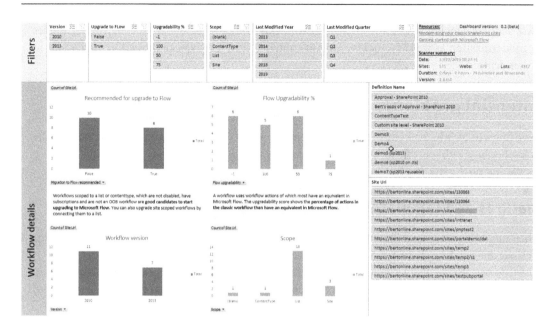

Figure 2.3 – A ModernizationWorkflowScanResults.csv file representation

Stream migration tool

This one is a bit more of a specialty case. Rather than focusing on documents, pages, and lists that we migrate from A to B, this tool (in preview at the time of this writing) gives us a way to transition from *classic Stream* to *modern Stream*. Stream is a cloud-only service that was the successor to the **Office 365 Video** service, but the backend hosting mechanism is changing. Microsoft is moving away from Stream having its own data store to saving all videos in SharePoint or OneDrive for Business.

Even the cloud is always changing. Just another reason why classic doesn't just mean on-premises and modern doesn't mean online. We're told that no retirement date from classic Stream storage is set yet. Feature parity is still lacking as things such as automatic transcription haven't made their way over yet. Microsoft indicates that once the Stream migration tool is released to the public, there will still be a 9-12 month transition period.

One of the driving factors of this decision has largely seemed to be compliance. Videos saved in Stream were not subject to retention or eDiscovery like other files in Office 365, so this effort is designed to address that. Metadata, transcription, and updated web parts to view video content will soon follow.

These additional tools give us some specialty instruments to scan for potential issues and opportunities for modernization as well as a tool for Stream content. Next, let's explore those potential issues in greater depth.

Potential problem spots

Some migrations inherently include content or features that cannot be migrated. These aren't really errors, because we expect them to break. We need to take note of a few items that will always cause trouble and pose a risk to a timely and successful migration.

Features and templates

When migrating from SharePoint Server to SharePoint Online, we've already seen that the worlds of classic and modern are very different. Of the five key modern experiences discussed in *Chapter 1, Classic versus Modern SharePoint*, site templates and pages have the biggest impact on the user experiences. Classic pages will need to be recreated. Classic site templates will need to be converted to modern ones, or at least have modern pages enabled.

Since classic site templates are really just blank sites with the right set of features enabled, it may be a more effective architecture approach to choose a modern site template for your target and choose the features you need. This may allow you to decide whether the current feature set is really necessary. Reducing reliance on classic features is a good practice today and may give you more future-proofing than migrating sites as they are.

From an architecture perspective then, it may be better to create target sites in a migration using one of the modern templates and only enabling the features you need to maintain legacy functionality (such as Document IDs or Document Sets, for example). Publishing infrastructure features should really be left behind whenever possible. Modern communication sites provide a greatly improved alternative to these features. Sections on modern pages are far more flexible than page layouts, and Power Automate features provide an improved experience for approvals.

Custom solutions

Solutions and features may cause the most trepidation to IT. This is one of the most significant planning challenges, in my opinion. Solutions in classic SharePoint over the years were packages of code-based artifacts, which could have been built using server-side .NET code or client-side script. **Full Trust Farm Solutions** were deployed as WSP files. The `Add-SPSolution -LiteralPath D:\Code\Solution_Name.wsp` command in the SharePoint Management Shell was useful for quickly deploying these solution files that could be easily produced with Visual Studio. Once the solution was added and activated, it could bring us web parts to add to a page, master pages to deploy across site collections, or farm-level changes.

So, why does this option no longer exist in SharePoint Online? A deployed solution is something that can have an impact on the entire farm when deployed by a SharePoint administrator. It could affect one or more web applications. An option on the deployment command could allow code-based resources to be installed in the **Global Assembly Cache (GAC)**, which would add components at a global level capable of impacting SharePoint and other services and products running the machine hosting SharePoint. Too much power in a package is too easy to create and implement, and certainly too big of a security risk to install in a cloud environment where we're renting space rather than buying it. SharePoint Online is by nature a multi-tenanted environment. So, any custom solution that is applied at the server level will affect not only our tenant but others as well.

Sandbox solutions were the answer to some of these concerns for a while (through SharePoint 2013). The solution was deployed by site collection owners rather than by a SharePoint administrator. A deployment was an upload of the solution file into a gallery within the site. This was supposed to put the power into the hands of the business stakeholders, although the solution still needed to be created by a professional developer. The code in a sandbox solution had access to a subset of the SharePoint object model, a subset of available .NET assemblies, and ran under a restricted Code Access Security policy. Usage limits and tighter monitoring added to the feeling of security, but in the cloud, this is still server-side code, and not safe for tenants sharing space.

The real question for planning is whether these custom solutions are really useful after a migration. Has the code within effectively been replaced by an out-of-the-box modern function? Does the solution deploy custom CSS, HTML, and JavaScript to fully rebrand SharePoint, which will no longer work on modern sites? Can a Power App or a flow in Power Automate give us an easier-to-maintain functionality that replaces the reason the solution was originally built? Would a third-party add-in help us achieve similar outcomes that we no longer have the burden of maintaining? One of the primary uses of solutions was to deploy classic web parts. Conversion of these web parts to modern **SharePoint Framework (SPFx)** web parts is a better approach and one we will discuss in *Chapter 3, Modern Options for Customizing SharePoint Online*.

Heavy branding

The use of custom master pages, CSS, or injected code that impacts an entire page is not going to work in modern pages. These items will usually migrate over, but remain only available in classic mode. Site designs and site themes will be discussed further in *Chapter 3, Modern Options for Customizing SharePoint Online*, and provide a modern alternative to these common classic approaches. Using the `AlternateCSSUrl` page property is not going to work on modern pages either.

Our assumption is likely to need to change as well. Instead of branding SharePoint like you would with a website made from scratch, we should treat it like any other productivity tool. We don't seek to brand Word or Excel. We might change the background color, but that's as far as we can go. We give up being able to specify pixel positions, exact margins and borders, and custom fonts for a responsive design that shrinks or grows to fit the device viewing our content. Out of the box with minimal customizations has become the standard for SharePoint Online sites.

InfoPath and SharePoint Designer

As noted in *Chapter 1*, InfoPath is on a self-destruct course. No reason to join the ride this close to the crash site! SharePoint Designer workflows will migrate over for the most part, as long as they are not based on the 2010 workflow engine. Those workflows will no longer run in the cloud and will need to be replaced. For out-of-the-box 2010 workflows, SPMT can help us migrate those automatically.

SharePoint 2013 workflows will continue to run for now, but we haven't been able to create new ones since November 2020. For anything we've built, we must look to rebuild using Power Automate. This may take time and lengthen the migration process, but it is a necessary and needed step on the path to living in the modern SharePoint world.

Any `.aspx` pages you've created using SharePoint Designer or system pages that you've customized will only work in classic mode. There is no path to modern with these pages and they should be seen as deprecated. I'll bet most of the reasons you chose that option in the first place has been included out of the box on modern pages.

Timing is everything – planning a successful migration

When we move from one house to another, the process can take a long time. Maybe we move all our stuff at one time or make smaller moves over a few days or weeks. There are some factors to consider when we look at the process of moving our content into SharePoint Online.

How big is the truck?

The size of the vehicle we use to move can make a substantial difference in how long it takes. As we noted earlier, Microsoft puts throttling limits in place to prevent an overload on your tenant's resources. We must essentially migrate through the same pipe that we use to conduct our daily business. Microsoft provides some guidance and estimates on performance and speed based on the types of content we're migrating:

> For light content migrations (ISO files and video files) – 10 TB/day
> For medium content migrations (list items and Office files ~1.5 MB) – 1 TB/day
> For heavy content migrations (list items with custom columns and small files ~50 KB) – 250 GB/day

Large files migrate faster than smaller ones. Migrating many small files can result in larger overhead and processing times. Files migrate faster than objects and list items. Microsoft also advises that *"Currently, the estimated upper limits for completing a Final Delta/Cutover event from Friday to Monday is approximately 100 to 150 million files or 12,000–15,000 users."*

So, it is not just the size of the truck, but what we're hauling in each load that matters. In my experience, expecting between 500 GB and 1 TB per day is normal. If the moving truck goes out in rush hour traffic, the move will take longer, so migrating overnight or at non-peak hours will help as well.

We must be realistic with the total amount of content to migrate. Most migrations of any size must take place over several days or weeks. That message should be communicated to users as far in advance as possible.

Options for efficiency

We usually have a handful of competing interests during a migration. How long will users have to work in two places? Will source data be read-only and for how long? Which users by role or necessity can't afford to lose access to files? When does access to the target location take effect? How can we preserve the sanity of the migrators while also providing a good experience to the business?

Here are some suggestions that have provided value to me over the years to balance these interests.

Choose to migrate by department or function. If we identify content that a group shares in our inventory, we can have several smaller migrations of their content without impacting other business units. This should include sites they access and their OneDrive for Business content if they use it to share with others in their area of focus. Several small migrations may make the overall timeline of a migration project take longer, but the impacts may be less significant.

Mimic a big bang by pre-migrating content. We can start with pre-migration before users even know data is being moved. That makes the migration time look short to the users, but also will result in more differentials to account for in the final push. This really only works in SharePoint, where we can control permissions, unlike OneDrive, where a user would be able to see and start interacting with their files immediately.

Move the big stuff first. Often, there are only a handful of people or business units who have an oversized amount of content. If we migrate that content first, we can usually script the later migrations in fewer batches with more content per batch.

Make a communications plan

Good communication with everyone involved is key to a successful migration. Impacted users should understand the overall timeline long before any work begins. They should have clear expectations as to what their experience will be before, during, and after the migration. They should understand how to validate their data post-migration and to whom issues or concerns should be addressed. The following high-level communication plan may help in your next migration:

One month – at least one month prior to migrations, send a *what to expect* message noting the timeline and place a heavy emphasis on the *why*. Users may have little disruption during the migration, but any disruption needs a rationale. A reminder to clean up old or unused files should happen here as well.

One week – one week before migration cutovers are to take place, remind impacted users of what to expect. This may be a single communication to be sent if we are going to do a single cutover, or multiple communications stretching out over time if we migrate departments on a staggered schedule.

The day before – send a reminder to complete the cleanup and to save all files in the source. Files in an open or locked state may not migrate and will require additional effort.

Day of migration – users will likely take comfort in getting a message when their content migrations are beginning and when they are finished. The latter should include specific testing and validation guidance. If source systems are going read-only, a reminder to that effect should be provided as well.

One week after – chances are that users will let us know whether there are issues, but it's possible that someone may be on vacation or just too busy to check and validate. Providing this check-in allows an opportunity for the users to feel valued and for us to provide some final validation to the work of migration.

There may certainly be other opportunities for communication. The old adage holds true with migrations. *You can't overcommunicate.*

Summary and planning document

In this chapter, we've defined common steps in the process of migrating content from file servers, SharePoint on-premises, and other cloud sources. We explored options to migrate content from various sources into SharePoint Online. We've seen that the source content somewhat dictates the tool that we use to perform the migration. For example, if we have a cloud source, we should first look at `Mover.io` as a tool. We also explored some common problems we might encounter when trying to migrate due to custom code, branding, or workflows.

At the close of each chapter, we offer a review of the planning questions important in the chapter and update the template to create your own SharePoint architect's planning and governance guide. In this chapter, we looked at the five elements important to every migration. We reviewed several tools to help us migrate this content from source to target and some common problems to expect along the way. Here are some notes to capture for your planning and governance documentation.

Content sources to migrate

Identify what needs to be migrated:

- What is the version of our SharePoint farm on-premises?
- What **cumulative updates** (**CUs**) have been applied to the SharePoint farm?
- What content should be migrated to SharePoint Online?
- Which sites are the most important from a business value or traffic perspective? These sites should have their pages modernized first, along with the home page of the intranet.

- Which sites have been abandoned, are outdated, or no longer serve a purpose. These sites should be deleted or at least excluded from migration.

- How many documents are still valid? Content owners should review their documents and delete whatever is possible.

- Is there file server content that needs to be migrated to the cloud?

- Do MySites need to be migrated to OneDrive for Business?

- Do personal drives need to be migrated to OneDrive for Business?

- Will we include personal files in the tool-based migrations or ask users to migrate their own content in a self-service style?

Migration tools

Identify which tools will be needed to conduct the process:

- Do we want to utilize tools from Microsoft or explore a third-party tool?

- Do we have the budget to purchase migration licenses?

- Are we migrating from one tenant to another?

- Are we migrating other cloud sources into Office 365?

- What level of reporting is desired once migrations are complete?

- What are our schedule and expected timeline for migration?

- Is Microsoft Teams a part of our migration plan?

Customizations to migrate

Identify problem spots that will not cleanly migrate:

- Which sites feature the highest degree of customization?

- Does the customization fulfill a business need?

- Does branding need to be retained?

- Do site owners know the capabilities of modern pages?

- Do we still have the source code for deployed solutions?

- Will we refactor a custom solution as a Power App or another modern alternative?

- Can our custom solutions be retired?

- Are there any business-critical workflows that need to be recreated quickly?

- Why were InfoPath forms initially created? Can they be recreated in Power Apps or Microsoft Forms?

In the next chapter, we will explore modern options to implement customizations in SharePoint Online. Most classic customizations won't cleanly migrate, so our best bet is to start with out-of-the-box functionality. When that doesn't quite meet our requirements, we need to explore the best practices for writing code that will work with modern sites and be as future-proof as possible.

Further reading

You can learn more about the Stream Migration Tool here: `https://docs.microsoft.com/en-us/`.

Modern Options for Customizing SharePoint Online

So far, we've compared SharePoint classic with modern and have seen that the move to the cloud has changed the approaches to customization and development that have worked in on-premises farms. We may see it as a limitation placed on code since our sites are hosted on shared servers in the cloud.

Another perspective is that SharePoint Online has moved to an *out-of-the-box-first* mindset. It's a productivity tool just like Word or PowerPoint rather than a development platform that must be customized to be useful. In any event, we can be assured that SharePoint Online is in a perpetually changing, evergreen state. The fewer customizations we add, the fewer changes we have to make when something breaks due to an unanticipated change in the underlying platform.

With that as our mindset, there may still be instances where we need to implement supported customization, organizational branding, or a function that doesn't exist unless we create it from scratch.

In this chapter, we will cover the following topics:

- What still works and what doesn't
- Site templates for reusable provisioning
- Implementing branding with site themes
- Formatting lists and adaptive cards with JSON
- Frameworks for development

The purpose of this chapter, like the rest of this book, is to assist with high-level planning. Let's start by looking at what works and what doesn't when it comes to customization and development in SharePoint Online.

What still works and what doesn't

I want to focus on the things we can and should be doing in this chapter rather than compiling an epitaph of lost functionality. However, this first section will help us understand what tools and approaches that we've relied on for on-premises and prior versions of SharePoint in the past are still viable today. Power Platform, a modern tool that offers low-code/no-code capabilities, can give us compelling alternatives to traditional development. We will discuss its role in depth in *Chapter 6*.

Perhaps we should start by making a distinction between two approaches for creating compelling and customized SharePoint Online sites. On one hand, we have what we'll refer to as *customizations*. These are not out-of-the-box properties per se and will require some declarative code such as **JavaScript Object Notation** (**JSON**) to be functional, but nothing must be compiled or deployed. Often, there are spots built into the existing UI where these customizations can be added, so the level of support from Microsoft is high. On the other hand, we may need to create things such as custom web parts or extensions with professional tools, which we'll refer to as *development*.

Let's start by listing the "what's in" and "what's out" regarding the development options. Then, we will move on to customizations.

Development

Custom development has long been a part of the story with SharePoint, but that story has twisted and turned over the years. There are certain development approaches that no longer work. Let's review them first:

- **Full-trust solutions**: Full-trust server-side code deployed in .wsp files is completely removed in SharePoint Online

- **Partial-trust solutions**: Sandbox solutions have been deprecated for a while now but are worth an honorable mention

- **Script Editor/Content Editor Web Parts**: No more injecting custom HTML, **Cascading Style Sheets** (**CSS**), and JavaScript directly into a page via out-of-the-box web parts on modern pages

- **Add-in/app part model**: Introduced with SharePoint Server 2013, this model essentially used iFrames, which are no longer as secure or as responsive as SharePoint Online demands

These development options did not vanish from the cloud; instead, they are now more focused on client-side execution (running in the browser rather than on the server) and supporting the responsive design of modern pages. Here are the primary development options we have for the modern SharePoint Online, which will be explained in greater depth before we finish this chapter:

- **Site templates and site scripts**: A reusable, JSON-based template file that creates a list of actions to run in sequence to create elements such as lists and fields. These are deployed to the tenant via PowerShell.

- **SharePoint Framework (SPFx) web parts**: Client-side web parts are built with your choice of JavaScript framework and deployed through the tenant app catalog. These can also be used to create single-part app pages and Microsoft Teams tabs.

- **SPFx Extensions**: While web parts are single components embedded in a modern page, extensions can be application or field customizers and command sets. An application customizer can target well-known HTML elements such as headers and footers. A field customizer can override the behavior of a list field as it is added or rendered. A command set can be used to add items to the menu of a list with custom capabilities.

- **Application programming interfaces (APIs)**: We can connect to SharePoint APIs to programmatically interact with data stored on our sites. The **Microsoft Graph API** can be leveraged to interact with a variety of Microsoft 365 tools and services and to create connections to external data. **MSGraphClient** is available to use within SPFx solutions. Alternatively, we can test **REST APIs** using the online **Graph Explorer**.

Customization

UI customizations have long been part of using and abusing SharePoint. For a time, heavy customization to prevent SharePoint from looking so much like SharePoint was the norm. Much of that functionality has been removed in modern pages or deprecated altogether. So, what's out? Let's take a look:

- **Custom master pages and page layouts**: This approach is still technically available in the cloud, but only for classic pages. Modern pages no longer use these features.

- **Custom CSS**: CSS sheets applied with the master page or at the site level is no longer supported with modern pages, nor can we inject CSS via modern web parts.

- **JSLink to customize fields and list views**: This is deprecated as it is not supported in the modern page framework.

- **Display templates**: This is a deprecated option as well, for customizing how search results are displayed. It only works with classic pages.

- **Snippets**: A gallery of HTML-based snippets that could impact a web part or control that is deprecated in modern pages.

Customizing sites, pages, and data in modern SharePoint is still very important, but the mechanisms to accomplish those tasks have changed. At the time of writing, we have the following options in place to customize modern pages:

- **Site themes**: Using the **Change the Look** option from the site settings gear icon allows us to pick a theme that has been defined by Microsoft as part of the platform. We can also pick custom themes under **Company themes**. A theme is a collection of colors defined in a JSON schema file. This may be deployed along with a site template or on its own via PowerShell.

- **List column and view formatting**: Text using a JSON schema can be added directly to the column of a list or library by choosing **Format this column** under **Column settings**. We can either use the built-in formats or go to **Advanced mode** and rely on many creative samples that the online community has created. A single field may be rendered or an entire list view.

- **Headers and Footers**: Each is a section under **Change the look** and allows us to use a predefined layout, along with images and selections from the color theme that have been deployed and used for the site. While this is a more out-of-the-box feature, it can have a large impact on the look and feel of the site.

Now that we've reviewed what's in and what's out, let's look at a supported and highly useful concept for creating consistent sites: the site template.

Site templates for reusable provisioning

Do you know what I love doing? The same thing over and over and over… said no one ever. Being able to create new sites from templates has been around for a long time in the SharePoint world. Creating sites from user-created templates is relatively new, however. For modern SharePoint sites, we can create a template file with JSON and deploy it using the **SharePoint Online Management Shell**.

The goal of using a site template is to create a reusable provisioning mechanism to enforce some measure of consistency when someone creates a new site. Microsoft provides a set of templates built into SharePoint so that after someone creates a new site, one of those templates may be chosen. You read that correctly. The template doesn't apply until after the site has been provisioned. On the plus side, that means an owner can assign a different template to a site at any future point as well, though anything created from a previous template remains and does not get deleted.

This is the experience the site creator will have. They will be prompted to select a template and will be able to choose from the Microsoft templates or from custom templates we create in the **From your organization** section, as shown in the following figure:

Start designing your site

Use pre-populated site templates that enable you to quickly create and customize engaging SharePoint sites. Browse templates or choose to design your own site.

Add or change your template anytime in Settings ⚙

Maybe later   Browse templates

Select a template

From Microsoft   From your organization

Crisis management
Share news, provide support, and connect people to resources in a crisis.

Department
Engage viewers with department news, events, and resources.

Leadership connection
Build organizational culture by connecting leadership and teams.

Learning central
Provide a landing experience for your organization's learning opportunities.

New employee onboarding
Guide new employees through your organization's onboarding process.

Showcase
Spotlight a product, event, or team using visual content.

Topic
Engage viewers with informative content like news and events.

Volunteer center
Onboard, train, and prepare volunteers for campaigns and events.

Figure 3.1 – Browsing a template when creating a new site

Many of the predefined templates are similar to what you would also find in the *Microsoft Lookbook* at `https://lookbook.microsoft.com`, which we introduced in *Chapter 1*. The big difference between templates and the LookBook is that anyone who can create sites can use the templates in SharePoint. Only global admins can deploy new sites from the Lookbook.

To create a new, custom site template, we need to begin by building the components that are needed in JSON. These are called site scripts.

Site scripts

Let's turn our attention to creating custom site templates. For this, we need two components – a site script and a site template/design. Starting with the site script, we have a set of actions defined by verbs in JSON that will be executed, in order, to build out the site components.

We can build out quite a bit of the structure and branding of a modern site by including the following in the JSON definition of the site script:

- **Content types**: Reusable sets of site columns
- **Site columns**: Reusable metadata fields that can be used across lists/libraries
- **List fields**: Metadata fields defined in a single list or library
- **Lists and libraries**: Content containers in SharePoint for items or documents
- **Views**: A configured set of viewable list columns defined to work together
- **Formatting**: For changing the default view of column data
- **Navigation links**: Links to content on the site or elsewhere
- A **site color theme**: Custom colors are normally found under **Change the Look**
- **Navigation layout (cascade or mega menu)**: One of the two modern layout options
- **Header layout and background**: You can select one of these options to change the layout or image
- **Site logo**: The image at the top left of a site
- **Joining a hub site**: The option for adding the site to a hub
- **Installation of an add-in from the app catalog**: For adding customizations
- **Feature activation**: For activating any required features for the site
- **Triggering a Power Automate flow**: An extensibility option for running additional actions
- **Adding people to groups**: To start setting up site permissions
- **Manage guest access**: For deciding whether outside users are allowed

The complete schema can be found here: https://docs.microsoft.com/en-us/sharepoint/dev/declarative-customization/site-design-json-schema.

Here is an example of a site script that includes both PowerShell and JSON. In this JSON sample, we are applying a theme, creating an FAQ list, adding a question field and answer field, and making changes to the navigation by removing documents and pages. At the end of the script, there are some commented lines. These allow us to add and apply the site design.

First, we need need to download the SharePoint Online Management Shell and use it to connect to our cloud instance. We can do so (as well as view how to get started) by going to https://docs.microsoft.com/en-us/powershell/sharepoint/sharepoint-online/connect-sharepoint-online. We need to pass the SharePoint admin center address as the connection point:

```
Connect-SPOService -Url https://<tenant>-admin.sharepoint.com
-Credential admin@<tenant>.onmicrosoft.com
$siteScript = @'
{
"$schema": "schema.json",
"actions": [
{
"verb": "applyTheme",
"themeName": "Sample Theme"
},
{
"verb": "createSPList",
"listName": "FAQ",
"templateType": 100,
"subactions": [
{
"verb": "addSPFieldXml",
"schemaXml": "<Field Type=\"Text\" DisplayName=\"Question\"
Required=\"TRUE\" EnforceUniqueValues=\"FALSE\"
Indexed=\"TRUE\" StaticName=\"Question\" Name=\"Question\" />",
"addToDefaultView": true
},
{
"verb": "addSPFieldXml",
"schemaXml": "<Field Type=\"Text\" DisplayName=\"Answer\"
Required=\"TRUE\" EnforceUniqueValues=\"FALSE\"
Indexed=\"TRUE\" StaticName=\"Answer\" Name=\"Answer\" />",
```

```
"addToDefaultView": true
}
]
},
{
"verb": "removeNavLink",
"displayName": "Documents",
"isWebRelative": true
},
{
"verb": "removeNavLink",
"displayName": "Pages",
"isWebRelative": true
}
],
"version": 1
}
'@
```

Now, we can run the necessary code to deploy the site script we just defined in JSON:

```
Add-SPOSiteScript -Title "SPO Sample Modern Site Script"
-Content $siteScript
#Resulting GUID from adding script is: 5ab1a5a1-6829-4c99-9032-
3273d408190b
#Set-SPOSiteScript -Identity 5ab1a5a1-6829-4c99-9032-
3273d408190b -Content $siteScript
```

The biggest piece that is missing from the list is the creation of a home page. This has been a common request from my consulting customers. It makes sense since the home page, including the location of the web parts and how they are configured, is how most people visualize the consistency of a template. Site scripts are more focused on the structural elements of a site, however, instead of the UI that's created inside of site pages.

Once our specific JSON for the site script has been created, we need to register it in SharePoint. We can do this using the REST API or with the `Add-SPOSiteScript` cmdlet in PowerShell, which looks something like this, where the `$siteScript` variable is equal to our JSON text:

```
C:\> Add-SPOSiteScript -Title "Our Custom Site Script 1"
-Content $siteScript -Description "Adds some consistent actions
to build out a templated site"
```

Using the REST API, we could use the JSON definition of the actions to execute, and first run, this command, supplying the title as text and the JSON variable just noted:

```
http://<siteurl>/_api/Microsoft.SharePoint.
Utilities.WebTemplateExtensions.SiteScriptUtility.
CreateSiteScript(Title=@title)?@title='Our test script'", JSON_
sitescript_variable)
```

Executing the script or the REST call will return an ID. We will use this ID in the next step.

Site template

The next step is to create the site design, also referred to as the site template. Looking at PowerShell, we see the need for a title, a description, the ID mentioned previously, and a web template value, which we can see in the following script example:

```
C:\> Add-SPOSiteDesign -Title "Demo Site Template" -WebTemplate
"68" -SiteScripts "<ID>" -Description "Our custom site
template"
```

The `WebTemplate` argument should match one of the built-in values to indicate the underlying SharePoint template for our new site. Even though we're creating a custom template, we are still just building on top of one of the following:

- **1**: Team site without an M365 group
- **64**: Modern Groups-connected Team site
- **68**: Communication site
- **69**: A channel site (used for Microsoft Teams private channels)

If we indicate that our template is for a channel site, the following actions will no longer function:

- Applying a theme
- Joining a hub
- Changing navigation links

- Adding users and groups
- Guest settings

Once the template has been registered with SharePoint, site creators and owners can see and select our template in the **From your organization** tab. This either shows up when the site is first created or if the owner goes to the **Site settings** gear and clicks **Apply a site template**.

If we want our custom template to be the default, instead of the predefined template from Microsoft, we can add the `-IsDefault` switch to the command. For communication sites, this would build on the **Topic** layout. For team sites, the default layout for the home page would be maintained. In either case, the site would be built using our custom list of actions in the associated site script.

Site scripts are limited to 30 actions if executed in PowerShell with the `Invoke-SPOSiteDesign` command. If executed by applying a template in the UI (or by running `Add-SPOSiteDesignTask`), that number increases to 300 actions or 100,000 total characters. We are limited to 100 site scripts and 100 site templates per tenant.

Template scoping

By default, a template is visible to everyone and available for them to use when creating sites. If we wish to limit which templates our users have access to, we can use scoping to grant rights to a template. Using the following PowerShell command, we can set the `-Principals` argument in the script to users or security groups. This can be done either at the time the template is registered or at a later point with the correct site design ID:

```
Grant-SPOSiteDesignRights -Identity <ID> -Principals
(groupName@tenantName.sharepoint.com, "userName@tenantName.
sharepoint.com") -Rights View
```

These rights can be revoked or viewed with additional commands, as defined here: https://docs.microsoft.com/en-us/sharepoint/dev/declarative-customization/site-design-scoping.

In addition to the behind-the-scenes structure that we can implement with a site template, we can also implement one of the most important branding aspects of a site: the color theme.

Implementing branding with site themes

While much of the company brand is implemented with the images you choose for pages and the verbiage used to write the brand story, brand colors also promote the corporate branding and often need to be enforced to maintain that standard across sites. Themes in modern SharePoint exclusively mean the color choices for site elements and do not include other elements, such as fonts. A theme can be deployed along with a site template or can be implemented separately.

As with site templates, there is a set of predefined themes (six light and two dark), each with a starting color palette. These themes may be used as offered or modified in the UI to change the preselected colors, as shown in the following screenshot:

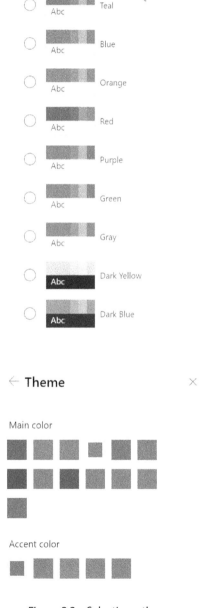

Figure 3.2 – Selecting a theme

The theme we select may include multiple colors that show up in the navigation, headers, footers, and web parts. If we choose the *Green* theme, for example, the background color of the pages will be white and the default color blocks for the hero, news, and quick links will be the primary green color we chose. The background color choices for the sections (none, neutral, soft, and strong) of a site page are also based on the colors defined in the template.

To create a custom theme, we can turn once again to JSON to define a dictionary of SharePoint color placeholders and the hexadecimal codes. We can start with one of the existing predefined themes and modify it to fit our needs or start from scratch. The full schema and the built-in themes are located here: `https://docs.microsoft.com/en-us/sharepoint/dev/declarative-customization/site-theming/sharepoint-site-theming-json-schema`.

The PowerShell `Add-SPOTheme` command can be used to deploy the JSON-formatted theme to our tenant. If we use the overwrite switch in the PowerShell code, we can make incremental changes to a deployed theme over time as well. It may take a moment for the theme to apply and users may need to refresh their browser for it to take effect. Here's an example of a theme being defined and deployed:

```
$palette = @{
"themePrimary" = "#1F3B6A";
"themeLighterAlt" = "#ffffff";
"themeLighter" = "#efc9df";
"themeLight" = "#e29ec5";
"themeTertiary" = "#128A00";
"themeSecondary" = "#C75000";
"themeDarkAlt" = "#1F3B6A";
"themeDark" = "#026610";
"themeDarker" = "#026610";
"neutralLighterAlt" = "#b0aead";
"neutralLighter" = "#F7F7F7";
"neutralLight" = "#F4C60A";
"neutralQuaternaryAlt" = "#9b9998";
"neutralQuaternary" = "#949291";
"neutralTertiaryAlt" = "#8e8d8b";
"neutralTertiary" = "#C75000";
"neutralSecondary" = "#b4b2b0";
"neutralPrimaryAlt" = "#2f2f2f";
"neutralPrimary" = "#1F3B6A";
"neutralDark" = "#30363B";
"black" = "#30363B";
```

```
"white" = "#FFFFFF";
}
Add-SPOTheme "Custom Org Themes" -Palette $palette -IsInverted
$false -Overwrite
```

Themes may also be applied through the CSOM API or the REST API. The REST API URL for theme management commands can use the following endpoints:

- `http://<site url>/_api/thememanager/AddTenantTheme`

- `http://<site url>/_api/thememanager/ApplyTheme`

The first endpoint would add the theme to the site. The second could take a JSON variable like the one we defined for PowerShell to apply to a particular site. More details are available from Microsoft here at `https://docs.microsoft.com/en-us/sharepoint/dev/declarative-customization/site-theming/sharepoint-site-theming-rest-api`.

We can hide the predefined themes with `Set-SPOHideDefaultThemes` so that only the theme(s) that resonate with corporate branding is used. `Remove-SPOTheme` may be used to remove a custom theme from the gallery if we change our minds.

One of the greatest challenges while creating a theme is knowing exactly where the colors will apply. Microsoft provides a theme designer, but it is more generic and geared toward Fluent UI design rather than being SharePoint specific. It allows us to use a color picker to apply the hex codes and can be exported to a JSON format. We can access that tool here: `https://fluentuipr.z22.web.core.windows.net/heads/master/theming-designer/index.html`.

One of the best resources I've seen is from Microsoft MVP Laura Kokkarinen, who has mapped each of the color tokens to the areas of a modern page where they surface. This can be found here: `https://laurakokkarinen.com/how-to-create-a-multicolored-theme-for-a-modern-sharepoint-online-site/`.

Themes have the most impact on site pages and global elements such as headers, footers, and navigation. While the background color defined in a theme may impact other types of pages, such as those generated for lists and libraries, the layout of data remains largely untouched by themes. For that, we need a different JSON schema.

Formatting lists and adaptive cards with JSON

Lists and libraries in SharePoint have long been akin to an Excel spreadsheet with information being displayed in rows and columns. Part of the built-in functionality allows us to sort, filter, and group data, but often, how the rows are displayed can be quite bland. It's not so much about making the list pretty as it is about adding visual interest to make the data easier to understand and consume.

In SharePoint Online, we can leverage JSON syntax and a formatting schema that can be defined and read by SharePoint. This format may apply to columns or list views with color-coding, layouts, and events such as button clicks or hover effects. Microsoft provides a handful of built-in options we can choose from to apply formatting. Alternatively, we can add the JSON text in **Advanced mode** to add our custom formatting.

Now, let's learn how to use this approach to format the look of a single column – that is, an entire list view – and briefly discuss how similar JSON formatting may be used with Adaptive Cards, which we will explore in more detail later in this chapter.

Column formatting

If we start with a list or library in a SharePoint site, we can click the carat drop-down to the right of a column name and click **Column settings | Format this column**. This will open a panel on the right-hand side that contains two sections – one to format the column and another to format the view. We can stay in this panel to format all available columns by using the **Choose Column** drop down:

Figure 3.3 – The formatting panel for a library with the modified column selected

Let's work through an example of using the **Modified** field shown in the preceding library view. The formatting can be accomplished by creating one or more elements with `elmType`. This is consistent with HTML, but limited to the following:

- `div`

- `span`

- `a`

- `img`

- `svg`

- `path`

- `button`

- `p`

- `filepreview` (a special type used to create a thumbnail/file type icon)

We accompany this with a set of attributes, styles, and rules. The **attributes** portion, defined for each `elmType`, mirrors HTML attributes such as `href` for an anchor tag. The **styles** portion allows us to add colors, text formatting, sizing, icons, and other forms of formatting that are consistent with name-value pairs found in **CSS**. We aren't using the CSS syntax, but we are using the relevant concepts. The **rules** portion allows us to apply the formatting conditionally, complete with operators and expressions.

Here is an example that illustrates what we're trying to accomplish. In this snippet, we can see that if the estimated airfare column has a value over $1,000, we want to change the look of the value so that the submitter of a travel request gets a visual indication that their costs may need to be reviewed for being excessive. The color of the text background will change, along with the color of the border. In a part of the code not shown, the icon has also been added to the left of the value:

```
"attributes": {"class": {"operator": ":","operands": [
         {"operator": "&&","operands": [
             {"operator": ">","operands":
["[$EstimatedAirfare]",500]},
             {"operator": ">","operands":
["[$EstimatedAirfare]",1000]}]
         },
         "sp-css-backgroundColor-BgGold sp-field-borderAllBold
sp-field-borderAllSolid sp-css-borderColor-DarkRedText","sp-
css-backgroundColor-BgLightGray"
         ]}}
```

This is the result in the list:

Airline ∨	Estimated airfare ∨
British Airways	ⓘ 1,200.00
Southwest	475.00

Figure 3.4 – The result of a formatted list column

A complete reference to the syntax elements can be found at `https://docs.microsoft.com/en-us/sharepoint/dev/declarative-customization/formatting-syntax-reference`.

We can either use **Design mode** to have the JSON written for us or **Advanced mode** to modify it directly. For example, if we want to format a date column named **Review Date** to show when documents are due for their annual review, we could add formatting to let us see past due items at a glance. If the date is in the past, we can highlight the item in red or increase the text size. If it is today's date, we can add an exclamation icon. If it is a future date, we can format the date value in green.

To do this in the **Design** view, we can select **Format dates** from the formatting panel and click **Edit styles** and navigate to **More styles**. We would end up with a designer surface that looks something like the following:

Figure 3.5 – Formatting a date column using the design view

What we end up with is a view where the **Review Date** column is formatted, but other columns are not, as shown here:

	Name ∨		Modified ∨	Modified By ∨	Review Date ∨	Add column ∨
⊡	Blog Post preview.docx		21 minutes ago	Nestor Wilke	September 1	
⊡	Contoso Purchasing Permissions.docx	⊖	21 minutes ago	Nestor Wilke	March 14, 2023	
⊡	Credit Cards.docx		21 minutes ago	Nestor Wilke	4 days ago ⚠	
⊡	Customer Accounts.docx		March 14	Lynne Robbins		

Figure 3.6 – Formatted column in the view

The other option would be to use conditional formatting written in JSON, in which case we could accomplish the same thing but have more control over the conditions used to produce the formatting by creating one or more rules. Either way, if we switch to **Advanced mode**, we can view or edit the JSON code directly.

The schema would contain something similar to the following at a minimum. We start with the schema element, which uses the Microsoft link and produces a starting element. For date fields, along with number, text, and choice types, we start formatting the output in `div`. Here, `txtContent` is outputting the value in the field using a placeholder:

```
{
  "$schema": "https://developer.microsoft.com/json-schemas/
sp/v2/column-formatting.schema.json", "elmType": "div",
"txtContent": "@currentField"
}
Excel-style conditional expressions will be added if we have
conditional formatting.  The biggest piece that is missing from
the list is the creation of a home page. This has been a common
request from my consulting customers. It makes sense, as the
home page, including the location of the web parts and how they
are configured, is how most people visualize the consistency
of a template. Site scripts are more focused on the structural
elements of a site, however, instead of the UI created inside
of site pages.
```

In addition to changes in format, we can also use column formatting to create interactive elements.

Advanced column formatting

Column formatting also provides a way to make interactive elements, such as clickable links, buttons, or ways to display column data on a hover action. For example, we could add a new column using the **Single line of text** data type and name the column `Delete`. Using the same process we've just gone through, we could format the column and add the following JSON in the advanced editor:

```
{
"$schema": "https://developer.microsoft.com/json-schemas/
sp/v2/column-formatting.schema.json", "elmType": "button",
"txtContent": "Delete me", "customRowAction": { "action":
"delete"}
}
```

`elmType` is set to `button`, which will display a button in every row under the new column we created. The button will have the word **Delete** on it and will invoke the delete file action with an accompanying prompt when clicked, as shown in the following screenshot:

Figure 3.7 – Using column formatting to create a functional Delete button

The `customRowAction` property of the button can also be used to add the following action types:

- `defaultClick`: This does whatever the default click would do on an item
- `share`: This opens the sharing dialog
- `editProps`: This opens the item properties panel in edit mode
- `openContextMenu`: This provides the same context menu you would see when clicking the dots to the right of the item or filename
- `setValue`: This modifies an item internally without showing an editor
- `executeFlow`: We can pass the ID of a Power Automate flow to execute when the button is clicked

Other advanced options include hover effects. When we hover over someone's name in a people column, a card pops up to show their profile information. We can add a similar capability to any column. For example, in the following screenshot, we have a button called **File Preview** that has been enabled to show us a preview of the file if we hover over the button in a particular row:

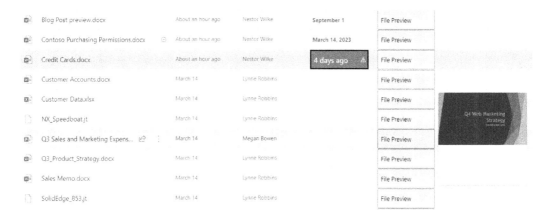

Figure 3.8 – Hovering over a button to see a file preview

We accomplished this with JSON column formatting, as shown in the following code block, with the `customCardProps` element allowing us to use `openOnEvent` to be associated with the hover action and `formatter` for our custom card to be an image that uses the built-in thumbnail token:

```
{
  "$schema": "https://developer.microsoft.com/json-schemas/sp/
column-formatting.schema.json",
  "elmType": "button", "txtContent": "File Preview",
"customCardProps": { "openOnEvent": "hover",
  "isBeakVisible": false, "formatter": { "elmType": "img",
"attributes": {"src": "@thumbnail.large"}}}
}
```

In addition to the changes we can display in a single column, we can use the same designer or advanced mode approaches that impact the entirety of a list view.

View formatting

SharePoint offers the built-in options of either an alternating row style or conditional formatting. With conditional formatting, we can either have multiple independent conditions or use and/or operators to connect conditions. When using multiple conditions, the order can make all the difference. For example, if we want the background color of the row to go red for items that are past due, but leave all others unformatted, we could add the following in the designer:

Figure 3.9 – Formatting the view with multiple conditions

If the order of the rules were switched, all rows before today would have a red background, even if there were no values in the **Review Date** field.

Formatting the entire list view opens up some amazing customization possibilities. By only using JSON formatting, we can format our list items in an accordion format as tiles, as task cards similar to Planner, as a board of sticky notes, formatted like a Twitter feed, and many more. A large list of samples that can be explored and implemented can be found in the following GitHub repository: https://github.com/pnp/List-Formatting/tree/master/view-samples.

Adaptive cards

An **adaptive card** is also a UI snippet created using JSON from a common Microsoft schema. This is a separate schema from view or column formatting, but it's worth mentioning, given how similar the approach is. More information is available at `https://adaptivecards.io/`, but for our purposes, we can define these cards as being built generically but rendered in a specific application. An adaptive card can be created using a code editor or Adaptive Cards online designer (`https://adaptivecards.io/designer/`). We may think of this conceptually as creating HTML that can be rendered using CSS in the browser.

The card contains the data elements we wish to display but ultimately gets rendered by a host application in such a way that it fits with its user interface. SharePoint Online uses these cards to build the dashboard for Viva Connections, which we will discuss in greater depth in *Chapter 7*. Additionally, Microsoft Teams chats, Power Virtual Agents and the Bot framework, Microsoft Outlook actionable messages, Power Automate, and custom renderers can make use of the same adaptive card.

Adaptive cards can be used to display data, including rich text and media, or to receive inputs. It also has the concept of actions, which act similarly to events. Adaptive Cards aren't available within the DoD environment (an elevated government environment for the Department of Defense), but we are seeing a broadening integration into other tools outside of SharePoint. It's safe to assume adaptive cards will very likely make their way into the modern SharePoint experience more broadly as time goes on.

For example, at the time of writing, we can post an adaptive card using an action in Power Automate flows to a private chat for a specific Microsoft Teams user or to a channel in a specific team. That card can be used to present information or to wait on the user to input information and submit it back. A deeper dive into the Teams integration can be found here: `https://docs.microsoft.com/en-us/`.

With the help of some JSON formatting, we can change the look of lists and libraries to make them more visually appealing and more functional. There are limits to what we can customize, however. To exceed those limits, we may find ourselves in need of custom development.

Frameworks for development

The options we've discussed so far have been one step away from implementing SharePoint with out-of-the-box functionality. Even though JSON and PowerShell have been involved, our approach can be better described as customization than custom development. However, we would be remiss if we didn't at least address some of the best-practice options, tools, and practices around developing components for SharePoint Online and programmatic ways to work with its features.

There are three primary areas I would like to focus on that represent a central core when planning to architect solutions inside of SharePoint and using it as a platform, rather than a data source – provisioning, UX extensibility, and scripting/APIs.

Provisioning

We have already explored the use of site templates as a mechanism for provisioning new sites and associated artifacts, but it is not the only option we have. The Microsoft 365 platform community exists as a repository of code artifacts, and the community members from Microsoft and partners who curate and maintain them. The site, which is located at `https://pnp.github.io/`, offers guidance, samples, blogs, tools, and recurring live calls and events to support development across SharePoint, Teams, Microsoft Graph, Adaptive Cards, and the Power Platform.

One of their initiatives is an open source provisioning engine that builds on and extends the capabilities of the site templates we've viewed so far. The **PnP Remote Provisioning Engine** is a tool that relies on PowerShell or C#. We will focus on the more common usage of the SharePoint PnP PowerShell module cmdlets.

We can begin with one of two types of templates – site or tenant. The *site* template is most similar to what we've explored so far but can be expressed in XML, JSON, or an Open XML format with a `.pnp` extension. The *tenant* template is an extended version that allows you to provision artifacts in other places such as Microsoft Teams or Azure AD.

To use the PnP engine, we can start by following these steps:

1. Creating a new site in SharePoint.
2. Saving that site as a template
3. Connecting to a specific site URL in PowerShell by using `Connect-PnPOnline`.
4. Executing the `Get-PnPSiteTemplate -Out "ourTemplate.xml"` cmdlet, which will result in a template file

The XML may be created manually as well using a schema, as shown here: `https://github.com/pnp/PnP-Provisioning-Schema`.

For more granular control over what is exported, we can configure the engine using a JSON schema, as shown here: `https://aka.ms/sppnp-extract-configuration-schema`. For example, we may only want to export certain lists, exclude permissions defined on the template site, or include terms from the term store. As noted earlier, site pages, including the home page, are not included as part of a JSON site template, but they can be included with the PnP provisioning engine.

When we are ready to apply the template to a site, we can run the following cmdlets:

```
Connect-PnPOnline -Url <complete path to our site>
Invoke-PnPSiteTemplate -Path "ourTemplate.xml"
```

This should result in the template you've defined now being applied and visible on your chosen site. Don't forget to refresh your browser if you don't see it right away.

UX extensibility

SPFx is our go-to solution for creating custom modern web parts and extensions for pages. Both of these items are considered UI extensibility in modern SharePoint. SPFx is also the recommended tool for customizing Microsoft Teams and Viva Connections. While our focus is on SharePoint Online, we can also use SPFx to customize SharePoint 2016 and 2019 on-premises and use web parts on classic and modern pages alike. Going to `https://docs.microsoft.com/en-us/sharepoint/dev/` is a great place to start exploring SPFx development.

SPFx is used to create client-side, responsive, framework-agnostic solutions. This means that SPFx developers will need to bring their JavaScript framework of choice, such as React, Knockout, or Angular to name a few.

An app catalog site is necessary as the deployment target for solution packages. SPFx solutions may also be published to the AppSource marketplace and the SharePoint Store if they are being developed for broader, public consumption.

Visual Studio Code, a free cross-platform IDE, is often the tool of choice for creating SPFx solutions. Instructions for setting up your development environment can be found at `https://docs.microsoft.com/en-us/sharepoint/dev/spfx/set-up-your-development-environment` and include the following development and build toolchain components:

- Installing Node.js
- Installing Gulp to build projects and package deployable solutions
- Installing Yeoman and the Yeoman SharePoint generator to build common boilerplate code and project structure
- Trusting a self-signed developer certificate for testing and deployment

It is also recommended that we sign up for the Microsoft 365 Developer Program (`https://developer.microsoft.com/office/dev-program`), which provides us with a developer/sandbox tenant for testing, complete with sample data. On that tenant, we can preview and test web parts without deploying them by using the hosted SharePoint Workbench at `https://<site name>/_layouts/workbench.aspx`.

Web parts

Using SPFx, we can develop modern web parts using our JavaScript framework and components from the Fluent UI to remain consistent with other tools in M365. We can interact with the page context using a built-in object and interact with SharePoint data through the SharePoint REST API.

For example, to get list data, we would use `https://<tenantName>.sharepoint.com/_api/web/lists`. To accommodate this communication, a helper class, `spHttpClient`, is available to execute requests against the API. We can use **Syntactically Awesome Style Sheets (Sass)**, a CSS pre-processor, to implement CSS in our custom web parts and output HTML to `div` on the target page using the `render` method.

Extensions

In addition to web parts that are self-contained within a page, we can also develop customizations for other aspects of the SharePoint UI. Extensions are built using essentially the same tool set as web parts (though we can't test them with the workbench) and rely on modern JavaScript frameworks for development. There are four types of client-side extensions we can create:

- **Field Customizers**: This is essentially a programmatic way to define a new list field type with its own formatting, similar to what we explored with JSON column formatting but while creating a new named type of the `list/library` column.

- **Application Customizers**: These custom components allow us to access and modify well-known page placeholders, so they are great for custom headers and footers.

- **ListView Command Sets**: We can use these to add new items to the toolbar menu of modern lists and libraries with an icon and text. Think of these as custom actions.

- **Form Customizers**: These are a programmatic way to override the new, edit, and display forms associated with a list or library. At the time of writing, this is in preview release.

PnP web parts

As an open source community, PnP offers many examples of custom SPFx projects, along with a client-side library that can assist with simplifying REST API calls within web part and extension development. A rich set of samples using a variety of JavaScript frameworks to inspire our creativity and provide a starting point for our development efforts can be found on the PnP site here: `https://github.com/pnp/sp-dev-fx-webparts/tree/main/samples`.

One particular PnP solution worthy of note is **PnP Modern Search**. The classic search experience included a site (and template to create more) called **Search Center**. This site includes web parts that can be used in the site itself, or we can enable features in other sites to use these search box, filter, and refinement web parts anywhere. These web parts are functional and out of the box but can be modified to allow us to customize the way search results look and how they can be filtered. The refinement panel web part is especially useful in being able to filter search results based on metadata and taxonomy values.

These web parts are no longer part of the out-of-the-box modern search experience, but this functionality can be restored with the use of the PnP search web parts. A search box, search filter, search results, and a search vertical web part are available when the solution is deployed and works on all modern pages. You will see the following view when adding the web parts:

Figure 3.10 – Adding the PnP search web parts to a page

Full documentation and the downloadable solution can be found at `https://microsoft-search.github.io/pnp-modern-search/usage/search-results/`.

APIs and scripting

Let's conclude this chapter by briefly looking at the options for APIs and scripting to connect to data and control SharePoint objects and data. We've already mentioned the `spHttpClient` object, which can be found within the SPFx, to make REST API calls. Here are some other API and scripting options at our disposal:

- **SharePoint REST API**: A service comparable to client object models that supports CRUD operations from SharePoint sites and solutions. The following tools commonly are used to interact with the APIs outside of code:

  - **PnPjs**: This is an open source JavaScript library for consuming SharePoint and M365 REST APIs: `https://pnp.github.io/pnpjs/`

  - **Postman**: This is a desktop application designed for testing APIs: `https://www.postman.com/`

- **SharePoint CSOM API**: The client-side object model available to .NET or JavaScript code.

- **Microsoft Graph API**: A REST API for accessing SharePoint and other Microsoft cloud services, including Teams, Yammer, To Do, Planner, and more. Outside of code, we have **Graph Explorer**, a web-based tool for testing Graph API endpoints: `https://developer.microsoft.com/en-us/graph/graph-explorer`.

- **CLI for Microsoft 365**: This is an open source solution capable of running on Windows, Mac, or Linux, similar to PowerShell: `https://github.com/pnp/cli-microsoft365`.

The developer story for SharePoint is still a strong one. Its out-of-the-box capabilities are richer than ever and Microsoft has provided robust options for customization using JSON. When we need to customize SharePoint even further, we can't use the same on-premises approaches of full-trust server-side code and full rebranding with master pages, but we can create custom web parts and work with a page placeholder model that is more consistent than relying on static page structures.

Summary and planning document

In this chapter, we've explored the best practices for customizing SharePoint Online. JSON has played a major role in how we can express site templates for reusability, site themes for colorful branding, and formatting list views and columns in creative and compelling ways. We saw how the open source patterns and practices community is supporting the efforts of developers by creating its own templating engine and useful search web parts. Finally, we explored the options for SharePoint custom development that leverage SPFx.

In the next chapter, we'll turn our attention to the role that M365 Groups play in modern SharePoint and as a foundational concept for collaboration and security in Microsoft 365.

Customization and custom development can take much effort and expense, but if done well, they can extend the usability and capability of an already solid platform. Let's look at some questions to ask when compiling the customizations section of our planning document.

Branding and templating

- What are the critical components of corporate branding for SharePoint sites?

- What are the four most important colors in our branding guide?

- What are the overall types of sites that we will create repeatedly?

- Which lists and libraries will be important for defining reuse across multiple sites?

- Which navigation elements should remain consistent across all newly created sites?

- Does the home page or other site pages need to be included in a site template?

Formatting

- Which lists would benefit from richer data formatting?

- Do certain lists lend themselves to color coding? If so, name them.

- Is a card, board, or another view more conducive for interpreting certain list items?

- Do power users know how to utilize predefined view formatting options?

Development and provisioning

- Are developers in the organization familiar with JavaScript frameworks? Which ones?

- Have developer tenants been created for the development team for testing?

- Do we need a process to provision Microsoft Teams artifacts along with SharePoint sites?

- Is there an opportunity to build an adaptive card that works in more than one place?

- Do we have custom data that requires a web part for rendering and interaction?

- Have we used the Content Editor or Script Editor web parts on classic pages and need a replacement?

- Do we have Power Automate flows that interact with list data and may need to be initiated directly from a list item?

Part 2: From Lone Wolf to Pack Leader – SPO Integrations with M365

This is the core section where all readers will understand that SharePoint is no longer a standalone server or platform. It is the backbone of document collaboration and information communication in M365. We will review how other tools integrate into SharePoint and how SharePoint provides them with file and data storage capabilities. We will review collaboration tools such as Teams and Yammer, automation tools in the Power Platform, and how the Viva platform enables SharePoint to drive corporate communications and knowledge management.

The following chapters are included in this part:

- *Chapter 4, Understanding M365 Groups as the Foundation of Collaboration*
- *Chapter 5, Magic Tool in the Toolbox – Integrating SPO with Other Collaboration Tools*
- *Chapter 6, Making SharePoint More POWERful*

4

Understanding M365 Groups as the Foundation of Collaboration

From a humble home to the mightiest castle, a structure is only as solid as its foundation. For homes, that may be stone, concrete, or the like. For collaboration in Microsoft 365, the foundation is the cloud-only concept of groups. These groups contain individuals who need to work together and own content together that may span multiple tools and systems.

Microsoft 365 groups are the glue and connective tissue to bring these multiple components into a single ecosystem, including SharePoint, Teams, Planner, Stream, Outlook, and connections to external data and software as well. While our goal in the next chapter is to look at what these other tools provide, in this chapter, we will explore the simple but powerful concept of M365 groups:

- The role of M365 groups in collaboration
- The provisioning and life cycle of groups
- Making good connections – Office 365 connectors
- Security and sensitivity

This chapter will help us to better understand the role of M365 groups as the underpinning for collaboration. We'll look at how to provision, manage, and maintain these groups as well as some options for securing them for the organization and guests.

The role of M365 groups in collaboration

In its purest form, an M365 group is similar to other groups in Azure AD. It is a set of people that can be attached to a set of resources. It provides a permission boundary that can determine access to our collaboration tools and data. Let's compare the different types of groups that we might find in the M365 admin center and what makes M365 groups unique:

- **Distribution groups**: A way to send email to multiple users at once. In SharePoint, we've been able to use distribution lists for audience targeting, but this has nothing to do with security.

- **Security groups**: The traditional **Active Directory (AD)** group, which may be synced to on-premises AD or exist in Azure AD. This is solely concerned with granting access to resources.

- **Mail-enabled security groups**: A mix of granting access and sending email to users.

- **M365 groups**: People who are identified as owners, members, or guests that are connected to the suite of resources provisioned when the group is created.

So, Office 365 groups are unique in that they serve a permission purpose, a communication purpose (as each group gets an email address), and can be used to connect to data and systems within Office 365 (such as Planner, Teams, and SharePoint) and outside of the M365 stack, such as connecting to an RSS feed in Teams or connecting a SharePoint team site to Trello to view tasks on a page.

Let's look at the composition of a group, including whether it is private or public, and whether it needs to be dynamic versus manually maintained.

Who's in the group?

An M365 group can only contain individuals, not other groups. Many organizations moving to the cloud have already spent a great deal of time configuring and maintaining security groups. The desire is to connect these security groups to resources such as Teams, only to find that M365 Groups is the only option. A distribution group can be upgraded to an M365 group, but in general, one group can't contain another group, so we often end up with duplication since different groups serve different purposes.

People in your organization who become part of an M365 group either become an owner or a member. Owners are not IT admins and don't automatically get access to an admin center, but they do have full control of the group itself and the resources it connects. You can have up to 100 owners per group, though the person who creates the group is the default owner.

Public or private

An M365 group can either be public or private. It is possible to start with one and then change to the other as needed. A private group is by invitation only. Owners can add members to the group from within the organization or invite guests from outside it with an email if the tenant is configured to allow it. More on that later. Members can essentially request for other people to join the group, but the owner is the gatekeeper, either allowing or denying those requests.

A public group is open for the entire organization to join, though no one is automatically added. Each user would have to opt in or join the group, but there are no locks on the door or guards to keep people away. Public groups are discoverable, and a person can be a member of up to 7,000 groups within a single tenant.

Dynamic groups

Some special cases for groups require an ever-changing assignment to better manage them as people come and go in the organization or move into or out of departments or roles. In Teams, we have the concept of an org-wide team that grabs all users in the Azure AD (even service accounts and guests). In Yammer, we have an *All Company* team that automatically contains all licensed users. For groups that need to have an automatically maintained member list, we can leverage dynamic groups. You can change a group's membership from static to dynamic (or vice versa) in Azure AD.

Dynamic membership for M365 Groups is based on rules with one or more expressions that essentially query users and add them to the group if they match, or remove them if they no longer fit the criteria. The properties that we can use to build the expressions include things such as departments, location data, and job titles. We can either use an exact match or a wildcard expression to build our groups and maintain them dynamically as well. This allows us to put marketing people in the marketing group or executives in a leadership group. This keeps us from having to entirely rebuild duplicates from all of our security groups. We can't, however, have dynamic groups built from the presence of a user in a pre-existing security group.

Congratulations! It's a group!

So, what exactly happens when a new group is born? A new group can be created in the M365 admin center, the Azure AD admin center, or in the applications that support and utilize them. By default, every licensed user can create a group on-demand to serve their needs. So, if a user gets a gleam in their eye thinking about bringing a new group into the world, they can go into Outlook and create a group. They can go into Teams and create a new team (which creates a new group first). They can create a new SharePoint team site or Planner plan or Yammer community.

Where we start from makes all the difference in which specific artifacts get created for that group, but there is always a common set that gets provisioned automatically when a new M365 group is created. The following visual shows the linked resources that are generated:

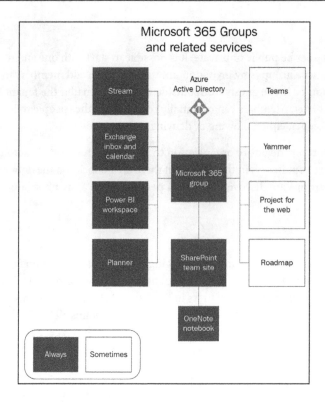

Figure 4.1 – Groups connected resources

The blue squares represent the gifts that are presented to our new group as soon as it enters the world. Groups always include a SharePoint site, Planner, a Power BI workspace, a OneNote notebook, plus an **Exchange Online** (**EXO**) mailbox and calendar. This is true regardless of where the group takes its first steps in life. Teams and Yammer are unique in that they rely on the basic set of resources but add their own features as well. Creating a team always means I get a group, but there are things such as chats and tabs that only live inside the Teams application. In Yammer, there are discussions, questions, and announcements that only live in Yammer, though a group is still created.

If a group is created without an attached Microsoft team, one can be added by one of the group owners from the lower left on the home page of the Team site, from the Microsoft Teams application, from the M365 admin center, or by an admin via PowerShell. The default in the M365 admin center is to create a team when creating a group.

If the group has over 10,000 members (first of all, yikes!), this can't be done. The following figure shows what an owner would see in Teams when creating a team from a group they already own. This is a process called **teamifying** your group, which is just fun to say, even if you don't need to do it:

Figure 4.2 – A screenshot from Teams when connecting to an existing M365 group

We see a clear picture that the group needs to always store files. It uses SharePoint for that. The group must rely on a way to communicate, which is with the EXO mailbox. The group may need to track assigned tasks, so there is always a Planner plan. Even if we don't think a component is necessary, we get it anyway. A past customer once told me that they wanted a group for task management only, not for files, so they deleted the SharePoint site. This promptly deleted the entire group. These resources are being provisioned because they're required for the group to do its job.

The Team site

The site that gets created to support any M365 group is a fully functional, modern site. The owners of the group are automatically added to the site owners SharePoint group with full control. The members are automatically in the site members SharePoint group with limited control, which equates to viewing and editing (the Edit permission level in SharePoint). This shows up as a single token in the members group, rather than an enumeration of each person.

Since there is no such thing as a visitor, or read-only user, in M365 Groups, we do have the option to just share the site with someone rather than adding them to the group (depicted in the following figure). Doing so would allow non-group members to interact with the site only, but nothing else that's attached to the group. We see that option in the UI as well as the option to add new people to the group via the site here:

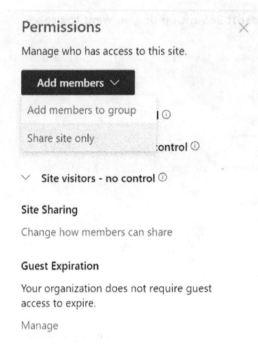

Figure 4.3 – A screenshot of the Permissions panel on a team site

Team sites that are groups-connected provide us with some context about the underlying group itself in the UI. Hovering over or clicking on the site title provides us with lots of information. We can view the group membership, which apps are connected to the group, the mailbox for the group, the option to invite or add others, leave the group, or opt in to follow the group by getting copied on emails it receives. The following figure shows us that UI:

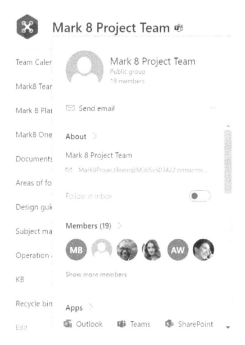

Figure 4.4 – A view of M365 group properties from a SharePoint site

This is a great way to see the little universe that is built on top of the group. It's a reminder that the SharePoint site doesn't stand on its own anymore. It is built on the foundation of the group, but it is also a resource of the group. It's a place to store pages and documents, but also just one piece.

The mailbox and calendar

As we've noted, the group always gets some EXO resources as well. Each mailbox must have a unique address. Even if the title of the group is the same as another, a string of letters and/or numbers is added to the email address. So, we might have a dozen groups whose title is HR, but only one mailbox with `hr@domain.com` as the address. The group mailbox is more like a shared mailbox than a distribution list. The mailbox, however, only allows email to be received. Neither owners nor members of the group have permission to send from the group mailbox. Being a part of the group will not automatically send you emails as well. You can either opt in to be copied in received messages or check the group's mailbox directly.

Having these resources seems to make perfect sense if we go into Outlook to create the group. Since we are already using a tool with emails as its currency, having this way to connect to the group is natural. The calendar is also a natural extension of the group. We're used to seeing calendars in Outlook that are stored in Exchange.

It's certainly possible that a user comfortable with and primarily working in Outlook could spend the bulk of their time interacting with the group from that application, rarely crossing into the others. At the top of the interface in Outlook for groups, we find buttons to link us to all the assets – email, calendar, files (which opens a SharePoint link), the notebook, and the settings menu for group management. There are also links to Planner and the SharePoint site itself rather than the document library opened by the files link. These links open a new browser tab and take the user out of the group interface itself.

For other groups-connected apps such as Teams and Yammer, having the email inbox makes much less sense. We have chats and discussions in those tools that are supposed to drive the interaction. As a result, the mailbox and calendar are essentially hidden. They certainly still exist, but there is no handy hyperlink waiting for us in the UI that takes us to the mailbox or displays the calendar. In fact, adding these can be more difficult than expected.

The **Calendar** icon in Teams shows us nothing about the group calendar. There is a channel calendar tab, but none for the group itself. The email is not automatically connected to a channel either. On the SharePoint site, we do have a group calendar web part but no way to view the inbox unless we click the link to Outlook under **Connected apps** and view the messages there. So, from a planning perspective, it may be safest to assume that the mailbox or calendar are *not* the best ways for a group to organize their communications or work.

Next, let's take a look at controlling the creation and life cycle of groups.

Provisioning and life cycle

Since groups are the foundation for all our collaboration tools and practices in M365, a question that quickly arises is who should be able to create them? Who should be responsible for making sure they are being created purposefully, managed well, and going away when their need has been fulfilled?

By default, all licensed M365 users can create a new group. This means that every person in your organization can create a group from Outlook, by creating a SharePoint site, spinning up a new plan in Planner, and all the other ways we've mentioned in this chapter. The default assumption from Microsoft is that adoption is strongest when people can get the collaboration spaces they need when they need them. It is the democratization of collaboration.

Everyone can create groups, but should they? Let's look at options for balancing adoption and strict governance. We'll see that good naming and expiration policies can make that job simpler as well.

Chaos and control

Good governance is all about balance – the balance between *chaos* and *control*. When it comes to M365 Groups, we can keep the Microsoft default and let people create whatever they want, or we can swing to the complete opposite side and lock things down tightly to IT only. A balance between the two hopefully drives both *adoption* and *cohesion*. While corporate culture factors in, I would dare say that the option that is favored often depends on the perspective someone represents – end user or IT admin.

Let's consider some positives associated with users being able to create whatever they need:

- I don't have to wait for IT to create something for me
- I can be responsive to the business needs of my area or department
- I can have a greater sense of ownership of my data and the tools that support it

Of course, there are some negatives as well, as follows:

- Sprawl and duplication of groups that represent the same or similar business needs
- Inconsistency with naming
- Some groups may be orphaned but not deleted

Too much control makes adoption suffer. Too much chaos and groups become a nightmare to manage. The balance between chaos and control needs to represent the best of both worlds. So, what options do we have to strike that balance?

Controlling chaos

The balanced approach to controlled chaos requires a solution that provides the following:

- Provisioning that is quick, but not wide open
- Naming that is consistent, but doesn't require manual compliance
- Life cycles that are consistent, but flexible

Let's assume that we don't want to leave group creation wide open. We could take the complete opposite approach and lock down group creation to only IT admins or other selected users that get added to a security group. To make that happen, we need to run some PowerShell, which can be found here: `https://docs.microsoft.com/en-us/microsoft-365/solutions/manage-creation-of-groups?view=o365-worldwide`. The script takes an Azure AD security group and sets that only included users can create M365 Groups through the applications directly.

An administrator going through the admin center will still be able to create as many groups as they need. If the admin wants to create groups in the applications as well (for example, creating a team from Microsoft Teams), they will need to be added to the group. Any user not in an admin role must be in the security group, or else the option will either disappear (as in Teams) or break (as in Planner) if M365 Groups creation is attempted.

We don't want to leave things wide open. We don't want to shut things completely down. This leads us to a menu of options to implement a governance approach to group creation and management that takes a middle ground.

Option 1 – The manual approach

We can leverage the **PowerShell lockdown approach** we just described and manually add more and more people over time. This is an option that allows us to slowly build adoption and self-service. Perhaps it makes sense to ask potential groups creators to take some owner training since they will be the owner of the group, the site, the team, and other resources. There would still be a burden on IT to manage the membership of the security group, but it would allow users in the business to own the creation and curation of the groups themselves.

Option 2 – The low-code approach

A more automated option would be to leverage Power Apps and Power Automate to build a low-code solution for requesting and approving the creation of M365 groups. This assumes that the work of option 1 has already been implemented, but we may want an easier way for users to make provisioning requests for new M365 groups or related artifacts.

We'll explore the tremendous potential of these tools in *Chapter 6, Making SharePoint More POWERful*. Suffice it to say for now that Power Apps could provide the frontend to make the request for a new group and Power Automate could control the approval and provisioning of the group. Any flow that is triggered from Power Apps runs in the context of the user. So, that user would need to have permission to create the group. This can be achieved by recording the request in a data source and having an event trigger the flow.

Since this could also be the case for the services that are built from M365 groups, we can also use this approach to help users avoid the *what to use when* problem. Rather than having a separate form for all the group-connected services, we could ask a couple of guiding questions and create the right artifacts on the backend. Do you need to store files? Do you prefer to work in Outlook? Would you like to have an ongoing chat as part of your collaboration? These questions could abstract things a bit from users and end up amplifying adoption in the process.

Option 3 – Open source solution

If we don't have the time or expertise to build our own low-code solution using the Power Platform, we can leverage one of the free solutions found on the web to do the job instead. I would recommend one of the following two, for their trustworthiness and functionality:

- First, we have what I'm calling a **training-based solution**. It's open source but curated by multiple Microsoft MVPs. It is built on the Power Platform, which leverages a chatbot, Power

Automate flow, adaptive cards, and a SharePoint list to contain the data all packaged up to be added as an app in Teams. The goal is to add people to the security group allowed to create teams, but only after they have taken sufficient training to equip them to become a group/team owner (and by virtue of the SharePoint site's connect, a site owner as well). The solution is detailed here: `https://docs.microsoft.com/en-us/microsoft-365/community/should-everyone-create-teams`.

- Second, we can leverage a low-code solution from the Office Developer GitHub repository. The *Request a Team* app template may be deployed, once again as a Teams app, to automate the provisioning of groups by creating a new Microsoft team. It likely goes without saying, but if you don't want a team associated with your group, this is not the option for you. This app will provide users with a way to build a new team from scratch or from a predefined template. Some basic validation checks the name and creates an approval task via an adaptive card. Once approved, the team is provisioned automatically. The solution is detailed here: `https://github.com/OfficeDev/microsoft-teams-apps-requestateam`.

Option 4 – A third-party solution

If our desire is to do more than simply build in some governance when a new group is created, we can leverage more robust, paid, third-party tools to assist us. There are two heavy hitters (at the time of this writing) that provide options for exactly that – Orchestry and ShareGate Apricot:

- **Orchestry** is a deployable Teams app that provides full life cycle management of Teams and SharePoint sites. The provisioning module abstracts some details from the user, referring to these as **workspaces** to ease the *which tool when* conversation. The workspaces can be customized and based on either a live template or one that is configured in the tool directly. Approvals, metadata, naming, and expiration may be applied. SharePoint components and features, including the provided custom web parts, can be added easily. The life cycle management tools identify unused workspaces, manage permissions, check guest access, and help with archival of sites and teams.

- **ShareGate Apricot** is a standalone tool that is bundled with ShareGate Desktop, which is primarily a migration tool. Apricot is geared toward automating, managing, and monitoring Teams. A daily scan produces reporting to identify inactive or orphaned teams, unknown guest access, or other issues. The option to keep or archive/delete teams, and apply custom metadata and information protection labels help govern teams. With the migration component available as well, we can move or relocate teams and channels to keep our groups-connected artifacts scalable and pliable for changing business needs.

Naming and expiration

Regardless of which tool or approach we take, we can rely on two features available in Azure AD **Premium P1** or **P2** to assist us with our groups' governance. A **naming** policy allows us to enforce a naming strategy for groups that are created by users from the workloads or applications directly. In other words, if a user creates a group by going into Microsoft Teams and creating a new team, the naming policy will automatically apply on the backend if we've defined it. Groups created from the admin centers would not have the policy applied.

Our naming policy may include prefixes or suffixes attached to group names. This could contain a string that is consistent, such as **user-created group** (**UCG**), or be based on a user attribute that is populated in Azure AD. Since the user who initiates the creation process is the owner, the data would be pulled from that user's properties. We could leverage this to include data such as department or location in the group name. Just make sure not to have a group name that is longer than 264 characters. Also, keep in mind that this naming is most beneficial when viewing groups in the **Global Address List** (**GAL**) or in the admin centers. The group alias, site title, and team title can be changed after the fact. This means that the title displayed in the UI may be different than the actual underlying globally unique name for the group.

A list of blocked words may be defined as well, which would prevent users from creating groups that contain those words. There are two use cases for this really. One is to prevent inappropriate words from being added to workplace artifacts. The other is to protect against the use of meaningful organizational terms such as department names, project or initiative keywords, a group for the CEO, and the like.

An expiration policy helps to prevent the limitless growth and sprawl of groups over time. This policy gives a chance to pester group owners on a regular basis if their group is unused. If someone visits a channel, interacts with a SharePoint document, or interacts with the group in Outlook, that is considered enough activity to keep the group going automatically. This comes in especially handy for users trying on groups, teams, or sites for size for the first time or for demo purposes – so lots of `Test1`, `Test2`, and `Test3` groups that were never intended to be kept.

Group owners are given the option to renew their group or to allow it to be deleted. Emails are sent to owners 30 days, 15 days, and 1 day prior to group expiration. Deletions are soft deletes that can be recovered within 30 days. Groups with no valid owners can get forwarded to a specific email address for review. This is usually a distribution list for the IT or governance teams to review. We can set the policy for all groups or for a selection of groups (up to 500). Unfortunately, we can't have multiple policies each applied to a different set of groups. If a retention policy is in place that would impact the group, the contents are stored in the retention container once the group is deleted and kept for the number of days defined in the retention policy.

Groups on their own are powerful, but we have a mechanism at the ready for extending them and allowing them to connect our group to data outside of its borders.

Making good connections – Office 365 connectors

Just as much as groups are the foundational elements underneath all our collaboration tools in M365, they are also the glue that connects groups to data and services. To borrow another building metaphor, groups are like the mortar that connects each brick. We've already started to see this with how a site, team, plan, email, OneNote notebook, and calendar are all joined together by the group and in support of the group. This is also true for how the group (and the people it represents) can connect to other data, apps, and services within and outside of the Microsoft cloud.

Connectors may be added through the tools that leverage them. Connectors are one way to connect our group to the world – either to bring data in or send data out. We can find a comprehensive list of these connectors, filterable by where we want to use them, in the AppSource site depicted in the following figure (`https://appsource.microsoft.com/en-us/marketplace/apps`):

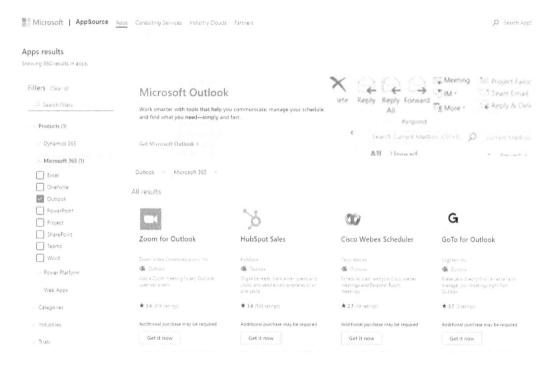

Figure 4.5 – A view of the AppSource site with Outlook selected as the product

Let's take a look at a couple of examples.

In Outlook desktop or the web, we can select a group and choose to add a connector from the store. The experience of the store is depicted here:

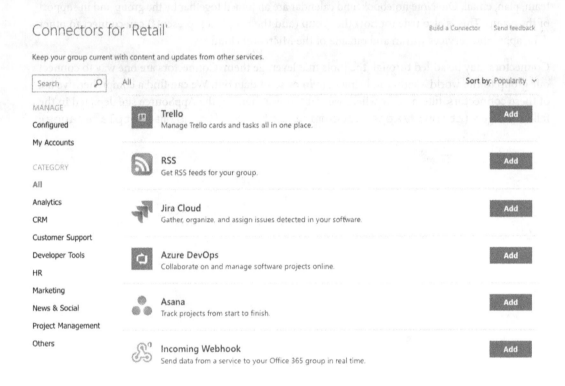

Figure 4.6 – Adding connectors from an Outlook group

We could add a Zoom connector to allow us to include a Zoom link directly in our meeting invitations. We could add a Slack connector to send email messages into a Slack channel. In Teams, we could add a connector for Trello as a tab to view our task board or add actionable chat messages that can build items directly from Teams. Additional information can be found here: https://support. microsoft.com/en-us/office/connect-apps-to-your-groups-in-outlook-ed0ce547-038f-4902-b9b3-9e518ae6fbab.

Connectors are also available for Power Apps and Power Automate. Each groups-connected tool can use connectors in ways that are a natural part of that tool's user interface. For SharePoint pages, these connectors can become part of the UI with a web part. Let's look at that in some additional detail.

The connector web part

For SharePoint sites, there are hundreds of connectors allowing us to connect to other services – some free and some subscription-based. The SharePoint UI is largely composed of web parts, however, so it makes sense that these connectors could be embedded on our modern pages. For team sites that are connected to an M365 group, we can use a single connector web part to bring in messages, visualizations, alerts, and other data.

By default, when we add the Office 365 Connectors web part, we get a single add button that opens the available list. Once the other service is selected, the web part property pane on the right will change to show the necessary configuration elements for that service. The following figure is an example of configuring the BingNews connector to deliver a digest of articles on the web related to change management directly on our SharePoint home page:

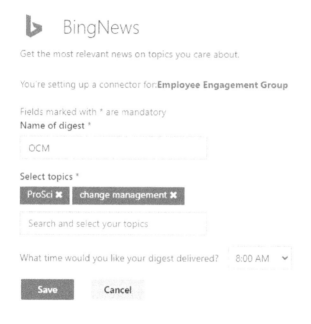

Figure 4.7 – The connector web part properties for BingNews

One of the most flexible connectors is the **Incoming Webhook**. Using this connector allows us to send data to our SharePoint site from web services that we develop or that are exposed from other systems that may not have a formal connector available. These messages are constructed with the same adaptive card JSON format that works in Teams, Outlook, and the Power Platform as well.

Security and sensitivity

M365 groups inherently give us a security boundary and a way to apply permissions for a set of users to a set of content items. In SharePoint, the members of an M365 group will always have Edit permission to the SharePoint site. That can't be changed. For anyone who isn't in our organization, planning for how those guests get access is crucial to maintaining the integrity of our data.

Guest access

Throughout this chapter, we've mentioned the possibility of inviting users from outside the organization into a group, site, or team. A **guest** is a type of user that doesn't have a licensed, regular account in your Azure AD. They will get an Azure **business-to-business** (**B2B**) or **business-to-consumer** (**B2C**) account, but won't consume a license with associated costs.

This is great for sharing content with external partners, vendors, and the like without having to pay to add them to your tenant. A guest can be added by an administrator in advance or in an ad hoc fashion by sharing content with them directly from SharePoint libraries. Owners of groups may add guests to their group by supplying their email from Outlook, SharePoint, or Teams. Members of groups may invite guests in Teams, but owners must approve the addition. Access requests may be enabled in SharePoint to accomplish the same. Guests access resources by clicking the link in the email they receive and creating the guest in Azure AD.

The problem with guest accounts in your tenant is the same problem you have with giving too many people the keys to your home. You're never quite sure when they're coming over, what closets and cabinets they're snooping in, or whether they've copied the key and given it to someone else. Once a guest is added in Azure AD, there is no great way to ensure we get their key back when their need for access is complete.

It's important to note that where we share things makes a big difference. Sharing a site or a document with someone would generate an email with a link, but they would only have access to the specific items shared. This would be like giving someone a code to enter your garage but not a key to get further inside. When a guest is added to a group or a team, they have access to almost all the content the group shares – chat messages in Teams and all pages or documents in SharePoint. This access would include read and write capabilities. The group calendar and email inbox would be off-limits, however.

Guest access can be disabled for groups altogether if desired by going to the global admin center and going to **Show all** | **Settings** | **Org settings** and selecting **Microsoft Groups** on the services tab. Additionally, we can decide to allow groups to have guests, but turn that feature off for Teams. This makes sense because of how easy it is for an owner to add a guest to a team and share so much with them automatically. It's also possible to block guests from specific groups.

We can limit how guests are added to the tenant. As noted earlier, admins may add guests directly to the tenant in the Azure AD admin center using *Azure B2B*. This allows the guest to be added with their name and email address along with a personal message and assignment to specific desired groups, as seen in the following figure. We can choose how much of the Azure AD resources a guest can see (such as other users or group membership) with options ranging from the same as a licensed user to nothing but their own profile:

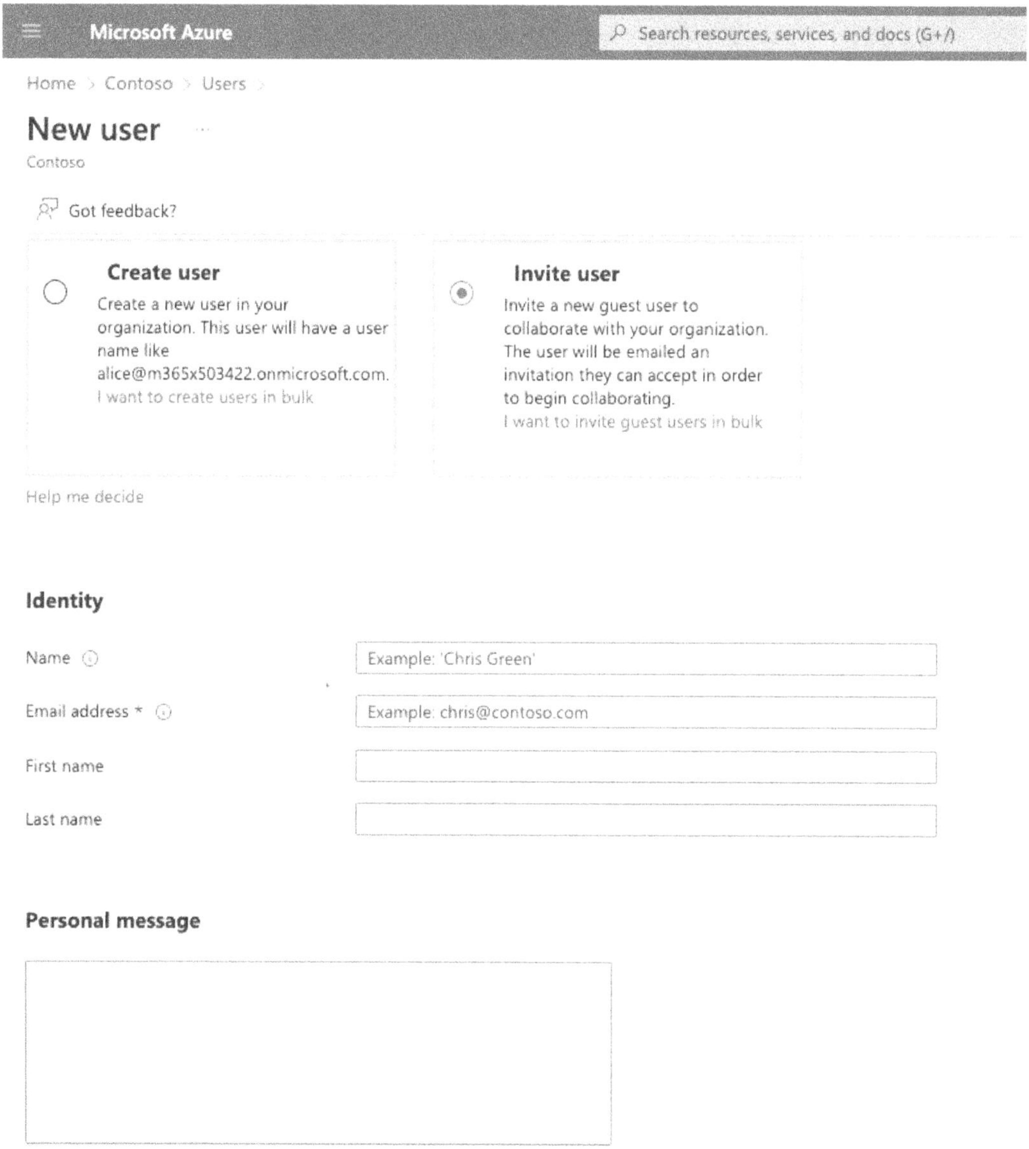

This is the next step after choosing to add a guest:

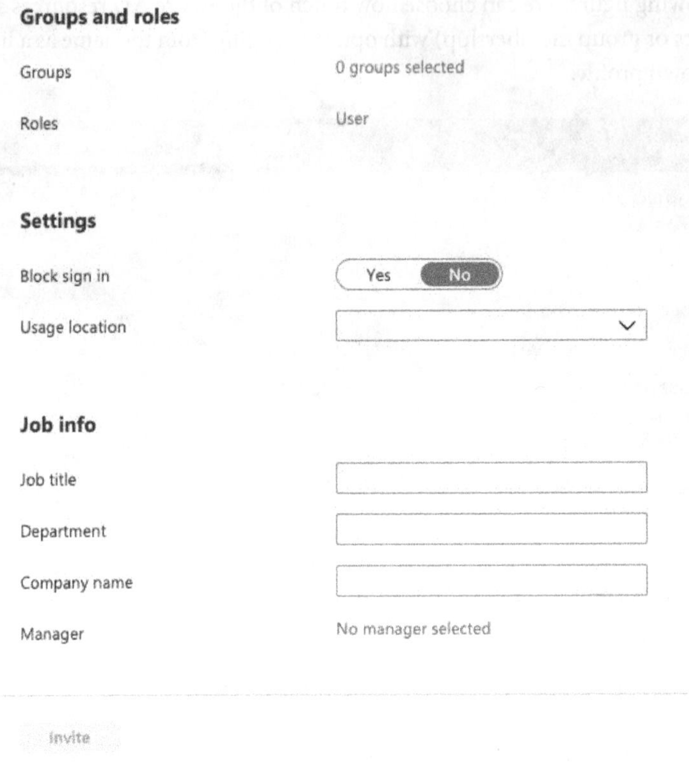

Groups and roles

Groups   0 groups selected

Roles   User

Settings

Block sign in   (Yes   No)

Usage location   [⌄]

Job info

Job title   [   ]

Department   [   ]

Company name   [   ]

Manager   No manager selected

[Invite]

Figure 4.8 – The admin center UI to invite a new guest

If we exclusively want to add guests in this way (and prevent them from being added by users), we need to log in to the Azure AD admin center as a global or security administrator and change the guest invite settings. There are four options:

- **Anyone in the organization including guests**: This is the default and means that your guests can invite other guests, so you're going to have a house full before you know it.

- **Member users and specific admins**: With this setting, a licensed user could add guests if they are a group owner, but guests could not add other guests.

- **Only users assigned to specific admin roles**: This is our option for preventing anyone other than admins from adding guests, either through the workloads or the admin centers.

- **No one can invite guests**: Shut the door, turn off the lights, and pretend that nobody's home. No external sharing at all.

The third option, **Only users assigned to specific admin roles**, does have a unique caveat. We can add users to a guest inviter role, which essentially makes them an admin for purposes of adding guests without giving them access to any other administrator content or sites. We also have the option to set collaboration restrictions that will either allow or deny guest invitations to specific domains. So, if we're working with a vendor on a project, we can make sure they can be invited to join us, but no one else.

Labeling our groups

Not all groups are designed to serve the same purpose. We've already seen how a group is a glue for our collaboration apps in M365. Some of those groups may be more sensitive than others. We must talk sweetly to them, avoid harsh chemicals, and turn the lights down low. Just kidding! The chats and files employed by the group may contain information that is sensitive such as proprietary business information, **personally identifiable information** (**PII**), or other data that needs special treatment.

We can use Azure Information Protection and Microsoft 365 Compliance Center to create and apply labels to content that identifies as a sensitive piece of content. Labels can be whatever makes sense for your organization, but are commonly something such as **Public**, **Confidential**, **Internal Only**, or **Top Secret**. When the label is applied, its configured protection settings are automatically applied to do things such as the following:

- Restrict content to users within the organization
- Decide who can read or who can print
- Dictate whether a message may be forwarded
- Set an expiration date on how long a document can be opened
- Define a custom header, footer, or watermark on a protected file

While the label may be applied by a user directly or automatically applied based on rules that we define (with an E5 license) to documents directly, we can also apply labels to the container itself. We can classify entire groups, sites, and teams. This only impacts the container, which means that it doesn't cascade down to the individual content items, though a single label could be applied to a site and a document within a library of that site.

Applied to a site or team, the label may help you to prevent external sharing for a site marked Top Secret or prevent guests from being added to a team marked Sensitive. Labels can be configured to work with **Conditional Access** policies to block site access from unmanaged devices. We can also use these labels to set the default sharing scope and link permissions when sharing content within a SharePoint site. The classification label for the container will display on a team or site and may be selected when that container is provisioned.

Summary and planning document

In this chapter, we've established the crucial, core nature of M365 groups. It gives us the connectivity and permission structure for sites, teams, communities, and more. Groups can be made with names of individuals added manually or dynamically. Since M365 groups can be created by any licensed user by default, we also explored the concept of governance. Provisioning, naming, expiration, and sensitivity labeling help our groups to work effectively and avoid redundancy.

At the close of each chapter, I want to summarize the information we've covered while also giving you a template to create your own SharePoint architect's planning and governance guide. This chapter has shown the importance of understanding M365 groups as the foundation for teams and sites as well as other tools in the M365 toolbox. Planning for group provisioning, governance, and life cycle will ensure a good balance between adoption and successful maintenance. Using groups as a means of connecting to data outside the Microsoft cloud allows all of our collaboration tools to be extended and to leverage other business-critical apps and data without leaving our M365 home.

Our planning document should include the following based on what we've discussed in this chapter:

- **Planning for groups**:
  - *Purpose*:
    - What do I intend to accomplish with the creation of a group? Email, permissions, or connectivity?
    - Are there already Azure AD security groups that may need to be duplicated or recreated?
    - Will the membership of my group be dynamic and automatically managed or static and manually managed?
  - *Which tool when?*
    - Do I need a group just to create an access control list?
    - Do I want to add real-time chat to my group that would require creating a team?
    - How frequently will my users need to interact with the shared email or calendar which may lead me to Outlook?
- **Governance**:
  - *Provisioning*:
    - Should we keep the Microsoft default of allowing all licensed users to create groups?
    - Does that decision apply to teams, communities, sites, and plans as well?
    - What naming strategy should we create for groups?
    - What is the minimum number of owners we should expect to have for each group?
    - What tools, solutions, or products should I use to assist me with automating provisioning?

- *Life cycle*:

  - What is our expiration policy? How long should a group be dormant before it is marked for deletion?

  - What should happen to the contents of a group if it is marked for deletion? Do we archive it?

  - Who will get notified if a group no longer has a valid owner?

- *Security and compliance*:

  - Should this group allow guests?

  - Should external sharing of content be allowed?

  - Will I need to use sensitivity labels for the group or its contents?

In the next chapter, we'll continue to explore the collaboration tools in more detail. If collaboration were a windmill, we've been focusing on the gear at the center with M365 groups. The arms of the windmill that really give it strength and purpose are the tools that connect the core of groups and SharePoint to power teams, communities, lists, forms, Stream, and task plans. The world of SharePoint just keeps getting bigger.

5

Integrating SharePoint Online and Other Collaboration Tools

I enjoy cooking. I have since I was a child, though I'm better at it now than I was then. Making something tasty that I enjoy that also makes others happy is very rewarding. To make a meal, there are only two required components – the ingredients and a heat source of some kind. You also need the skills to bring those two together successfully, which is why you're reading this chapter right now.

The process of cooking would be less enjoyable and the results less desirable if the only tools I had were my hands. I rely on several tools coming together, each playing a role in successfully preparing a meal. I need knives, spoons, spatulas, pots, and pans. I might also use a stand mixer, a spice grinder, or a steamer basket. Each tool adds its own unique value and I quickly come to rely on them all being available.

SharePoint has been known as a document management system or intranet tool for years. In the cloud, it still has a core set of functions it does well, but it is not the only tool in the toolbox. Some functions it used to provide, such as surveys, are deprecated. Other features, such as lists, have become a service of their own. Functions such as task management and tracking have moved to their own service as well. In this chapter, we will look at how SharePoint is not only a powerful tool in the toolbox but also the core of a set of tools that come together to *cook up something hot and tasty* in M365.

Our ingredient list for this chapter is SharePoint as the master sauce for making teams and communities, lists as the flavorful drops of data we can get to from SharePoint or via the Lists app, and we'll drop in a little Forms, Stream, and Planner to taste:

- SharePoint behind every Microsoft team and Yammer community
- Using Microsoft Lists to store and view data
- Using Forms for surveys and polls

- SharePoint, Stream, and video files
- Making plans and managing tasks

By the end of the chapter, our plan is to understand how SharePoint is the foundation of many M365 services, but also a place where those services can also shine and provide support to the core collaboration tool.

SharePoint behind every team and Yammer community

In the previous chapter, we discussed how M365 Groups provides the foundation for many connected services in the cloud. SharePoint Online sites are always created whenever a Groups-connected artifact is created. SharePoint becomes the place where files are stored for these services. Nowhere is this more visible or more important perhaps, than two of the other key collaboration tools in M365 – **Microsoft Teams** and **Yammer**.

Teams and SharePoint

Microsoft Teams is a chat-forward collaboration tool with a desktop, web, and mobile client experience that leverages storage in both Exchange Online and SharePoint Online. All chat messages are stored in mailboxes. If we have a *1:n* private chat or private chat group, each individual message is stored in a hidden folder in the mailbox of the chat participants. Any file shared is uploaded to the sharer's OneDrive for Business location. If there is a chat within the **Posts** area of a team, those are stored in the mailbox for the group. All documents are stored in SharePoint. Each channel has a **Files** tab, which displays the set of files stored for that channel.

Each time a team is created, the SharePoint site serving the M365 group is created, and the Documents library is connected directly to that team – one team to one site to one library. Within that library, a folder is created automatically for each channel that is created in the team. Folders that are created directly within the document library don't automatically create new channels, so the obvious intent is that Teams becomes the frontend experience for the user with SharePoint there to support it. It's also worth pointing out that an additional SharePoint site is created for each private channel in a team, of which we can have 30 per team.

SharePoint sites can still have as many libraries as we need, however. Only one gets a direct connection to the structure of the team. To make additional folders or libraries visible to team members, we can either jump over to the site directly or leverage tabs in Teams channels. The link to the site is always visible on the **Files** tab menu, on the right-hand side, labeled **Open** in SharePoint. This takes us into the folder mapped to the channel, but we can use SharePoint navigation from there. The tab options are as follows:

- **Website**: This tab allows us to enter virtually any URL as long as it starts with `https`. Using this option, we can display any view from any library or link to a specific folder.

- **Document Library**: Here, we can either connect to relevant sites (the four sites used last or most often are likeliest to display here or we can enter a URL just like with the **Website** tab).

- **SharePoint**: This tab is more flexible still as we can add a page, list, or document library from the same site that is linked to the team, or any SharePoint site we have access to. The latter option is essentially the same as the **Website** tab because we must supply the URL. In the following figure, we see the options for adding a tab to display this content:

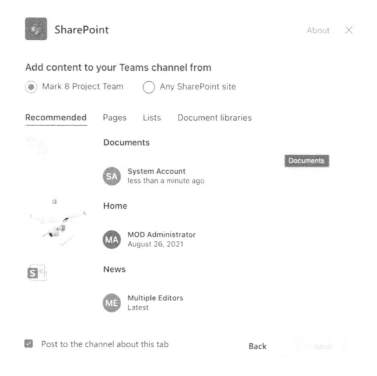

Figure 5.1 – The UI for the SharePoint tab in Teams

Having ready access to your document content in Teams is very important, but there is one more advantage of Teams and SharePoint working together. In SharePoint on-premises, there was a feature that allowed us to configure libraries to receive mail, allowing attachments to be saved as documents. For a while, this feature disappeared. With Teams, we now have an email address for each channel, which allows us to send messages with attachments and have those added as channel messages with attachments automatically being added to the SharePoint library and exposed in the **Files** tab. We can then continue the chat interaction in Teams as we collaborate on the files. Since this tight integration between channels and files exists, there are also some pitfalls to be mindful of.

Potential pitfalls

What if I delete a channel? What happens to the files? The channel is removed from the list on the Teams side but the folder in the library remains. If you try to create a new channel with the same name, however, you'll see the following message:

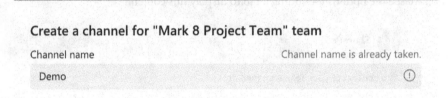

Figure 5.2 – Teams message when trying to reuse a channel name

For 30 days, that channel will remain available. A team owner can go into **Manage Team** on the ellipses next to a team name, to the **Channels** tab, and select **Restore**. The conversations will be rehydrated and the link to the library will be restored.

What if I change the name of the folder in SharePoint? This could be a common problem since there is a link between the channel and the folder, but any member can rename a folder in SharePoint. This action will not rename the channel in Teams. It will still exist, and conversations will be untouched. However, when we click on the **Files** tab, we'll see what I lovingly refer to as the *bicycle of doom*, the message as follows:

Figure 5.3 – An error message in Microsoft Teams

Changing the folder name back to the name of the channel should restore everything and resolve the issue. What if we look at the issue from the other direction, however, and rename the channel in Teams? In this case, the files will still display without issue in Teams. In SharePoint, the folder name will still keep its original name. This can cause confusion if some users navigate your content in SharePoint, while others go through the Teams UI. Preventing that confusion isn't easy though, as the rename command is no longer visible on that folder in SharePoint (you can still rename folders not connected to a channel though).

Yammer communities

Yammer is a communication and social networking tool for web and mobile that comprises conversations and files. Multiple discussion types such as announcements, questions, polls, and praise create a more robust way to communicate within a private or public community than Teams channel messages and chat. Yammer has been around for a while and has gone through quite a few technical architecture changes in recent years.

Today, modern Yammer is an M365 Groups-connected service just like Teams. As a matter of fact, the two services compete for the same set of groups. You can't connect a group to both Yammer and Teams at the same time – only one or the other. Since Yammer connects to a group, that instantly means that we're going to get a SharePoint Team site to store our files. Even though those files get stored in SharePoint, we can still use the Yammer UI to interact with them (seen here):

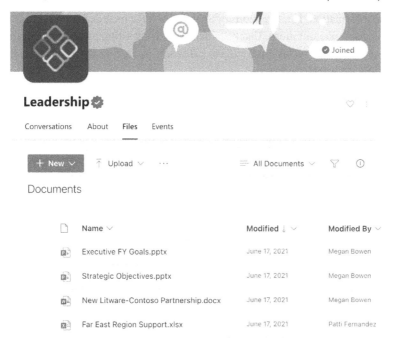

Figure 5.4 – A view of files in the Yammer UI

The **Files** menu opens a screen that displays the documents stored on the SharePoint site. In a similar manner to Teams, this reveals a folder in the **Documents** library. The path is `Apps/Yammer` and gets automatically created and connected when the community is built. Only documents in this folder will appear in Yammer. Any files that we add as attachments to discussions will also get added to that folder automatically.

We clearly see that SharePoint plays an important role in the structure of a team or community by providing a place to store files and a place to store pages. The latter can provide us with a richer UI to connect links, text, images, videos, and other content in a single location. Teams and Yammer both have their own organizing structure, which can sometimes feel fractured, so SharePoint pages can still be very helpful, even if the *front door* to collaboration is not SharePoint directly. Now, let's take a look at how list data in SharePoint is driving a new tool in the enterprise and even at home.

Using Microsoft Lists to store and view data

Our next tool to help us make a savory and desirable solution is **Lists**. SharePoint lists have long provided a mechanism for data storage. They're reminiscent of Excel spreadsheets in the cloud, with rows and columns of data. We know that lists and libraries are functionally similar behind the scenes. Libraries start with a document and can have additional metadata columns to further describe it. Lists are a collection of metadata columns that may also have optional documents as attachments.

List views have allowed us to sort, filter, and group data by this metadata, providing several ways to best consume the information. Prior versions of SharePoint have included special kinds of lists such as calendars, contacts, tasks, and more. That provided us with a starting point for applicable metadata by provisioning specific fields for us. Lists such as calendars also come with their own special views of visual data in specific ways.

Generally, if we've wanted to interact with the data in a list, we've needed to first go to the SharePoint site that contains it. Assuming we have permission to do so, we can find the list and view, edit, update, or delete its contents. While we can create views of the data that are private for just one person to see, most views are public, as is the data inside the list.

Microsoft Lists is an app in the M365 app launcher, just like Teams and SharePoint, which provides a unifying UI around multiple lists that we as users connect and interact with. It is not a separate storage location but a potentially better way to get to list data quickly and easily. You can create lists in Microsoft SharePoint, the Lists app in Microsoft 365, or Teams. The Lists app can also be installed as a progressive web app.

Wherever you start, the list will either end up in a SharePoint site if it is shared or in your OneDrive if it is part of *my lists*. Since OneDrive for Business is built on SharePoint, lists are stored on a site either way. We can view the UI that greets us when opening the Lists app here:

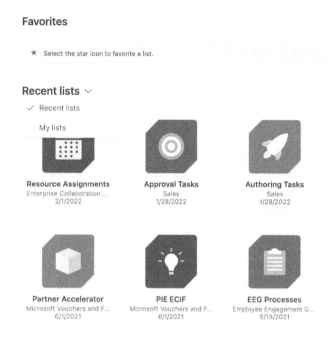

Figure 5.5 – An example of the landing page for Microsoft Lists

In the preceding screenshot, the lists are shown in one spot, but each one is actually residing in different SharePoint Online sites. If we click on the list from the Microsoft Lists landing page, however, we actually remain in that UI, rather than viewing a list within the context of the containing site. Just above the list title, there is a hyperlink to the underlying site. Clicking there takes us to the home page of the site. If we look for the list in the site contents, it will then open up with the site's theme, rather than the lists.

This represents two completely different ways for users to access the list data they need. If their work is in a handful of sites (think of people in human resources only working in the HR or benefits site), then it would make sense to have links to the lists on the home page for quick access. If someone interacts with lists coming from several sites and business units across the organization (think of a project manager who is getting data from multiple project sites), then it may save time to have all of those lists in one spot (possibly marked as favorites) on the Lists landing page.

Let's explore how and where to create a list, options for building different views of the data, and how to customize the views and data entry forms for the lists with a new JSON formatting standard.

Creating a list

When creating a list from Lists, SharePoint, or Teams, we are greeted with a screen that allows us to create a list from scratch, build one from Excel or another list, or we can select a template that will come with prepopulated metadata columns and some ready-made formatting. In the following screenshot, we see creating a list from the **Event itinerary** template. We can choose a color and icon to show up when using the Lists app:

Figure 5.6 – The UI of Lists after choosing a template

If we choose **My lists** as the save location, the list will be stored in our OneDrive for Business site, and we'll need to share it with other people if and when the time comes to do so. Saving to SharePoint automatically shares the list with those who can access the site.

The templates are limited but may provide some *what's possible* thinking if lists are new to you. It is probably worth noting that in order to create a list from Excel, you'll need to have the data in a table, not just a worksheet. One table may be selected to generate a Microsoft list. Whether we create a list from scratch or from a source, the option to add visual formatting to the list is a significant contributor to usability and adoption.

The availability of **My lists** and specific templates may be disabled if desired by using PowerShell as noted here: `Control Microsoft Lists - SharePoint in Microsoft 365 | Microsoft Docs`.

List views

The building blocks of lists are columns and rows. Each column is a piece of metadata containing a value that explains and relates to the list's purpose as a whole. For example, in the **Event itinerary** list we started previously, the template starts with columns such as session name, session type, speaker(s), and start date and time. Each column has a data type:

- **Session name** is a text column and is marked as required.

- **Session type** is a choice field providing a drop-down selector of values such as keynote, breakout, and workshop.

- **Speaker(s)** is a people field where one or more names or emails may be entered and resolved to a person known to Azure AD and SharePoint Online.

- **Start date and time** is a DateTime field. We have two data entry areas – one for the date driven by a calendar control and another selector for the time in half-hour increments.

We can add any new columns that are needed in addition to the ones provided by the template. The type of the column begins to define the UI when we add new items or edit existing ones. The details panel on the right side, or the entire screen showing the list in an editable grid, is where we see that interaction.

Views are the mechanism by which we see the presentation of the list columns as a unit. There is always one default view, but then additional views can present a subset of the columns of the list or provide a way to sort, filter, group, and/or total columns or groupings of columns. Views also provide a way to limit what is retrieved so SharePoint can work with large lists or lists with a large number of **Lookup** columns. The following figure shows the options for viewing a list. **List**, **Compact List**, and **Gallery** are predefined options that allow us to see primarily text or images when viewing the data:

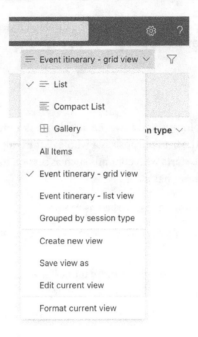

Figure 5.7 – Viewing the views

Four views are defined in the figure, but additional ones could be added as well. Creating a new view is as easy as clicking on the column headers, applying our change, and then saving the view (overwriting or as a new view with a different name) or by clicking **Create new view**. Here, we get four choices:

- **List view** – A tabular view in keeping with our discussion so far
- **Calendar view** – As long as we have a start and end date available for each item, it can be displayed as a calendar
- **Gallery view** – Perhaps more useful for libraries, but will allow us to see a view where an image is prominent, followed by other list data on a card view
- **Board view** – Very similar to Planner, showing a board view with different containers that list items can be dragged into

List formatting

The lists service has brought easy but powerful visual customizations for our lists, allowing us to build from what Microsoft provides, but creating our own formatting and enhanced visualizations. To accomplish this, let's explore list formatting for entire views in both the **List** and **Gallery** layout, and individual column formatting.

At the bottom of *Figure 5.7*, we saw the menu item to format the current view. Clicking there opens the details panel, which begins with a designer for views and columns. For list views, we can select alternating row styles to format the view table or conditional formatting, which allows us to provide color-coding to rows based on rules. In the following figure, we see that the row is highlighted in red if the capacity is over 100. This is in addition to formatting in the columns:

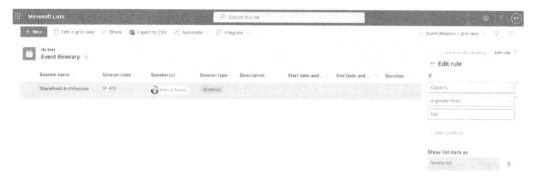

Figure 5.8 – A formatted list view

The card designer is available for Gallery layouts. We are presented with a checkbox list of all columns and can decide which are visible and the order in which they appear. We can choose whether the column title appears as a label above the text. For columns, we can choose from a drop-down list of formattable columns. The formatting options for a column are determined by its data type. For choice fields, for example, we have the ability to apply a background color to the cell, a colorful *pill* around the data (as we see in *Figure 5.8* for **Session type**), and a conditional formatting option that works like the previous option seen with view rules, except applying formatting to the column only.

While this is a powerful way to visualize data in Lists, we can also choose the advanced mode in each scenario. This opens up exciting possibilities to create our own visualizations from a standardized JSON schema. An example of that is seen here:

Choose Column

Session type	∨

Change the display of this column by adding JSON below. Remove the text from the box to clear the custom formatting. Learn more

```
 1   {
 2     "$schema": "https://developer.microsoft.com/json-schemas/sp/v2/column-formatting.schema.json",
 3     "elmType": "div",
 4     "children": [
 5       {
 6         "elmType": "div",
 7         "style": {
 8           "display": "flex",
 9           "box-sizing": "border-box",
10           "align-items": "center",
11           "min-height": "24px",
12           "padding": "5px 8px 5px 8px",
13           "border-radius": "16px",
14           "background-color": {
15             "operator": ":",
16             "operands": [
17               {
18                 "operands": [
19                   {
20                     "operands": [
21                       "@currentField"
22                     ],
23                     "operator": "toLowerCase"
24                   },
25                   {
26                     "operands": [
27                       "Meal"
```

Figure 5.9 – Viewing the advanced mode JSON editor for list formatting

We can view guidance on this JSON standard and see examples by visiting this page: https://docs.microsoft.com/en-us/sharepoint/dev/declarative-customization/view-formatting. There, we find some excellent documentation and a link to the GitHub repository where several compelling examples may be found. So, we see that SharePoint lists have come a long way over the years. The possibility of formatting data in columns and views to create rich, compelling user experiences is a better, more scalable, and simpler approach than custom development may have afforded us in the past. Let's turn our attention to the one component of lists we haven't discussed so far, New and Edit item forms.

Data entry

What about the data entry forms? In SharePoint on-premises, it was very common to leverage InfoPath or to customize the list form pages with SharePoint Designer. While both tools still work with SharePoint Online, both are deprecated, and neither work with modern lists and libraries. We'll discuss another option shortly, but in keeping with formatting options that SharePoint provides directly, we can determine which columns are visible or use the JSON formatting discussed earlier to change the header, body, and/or footer of the list data entry forms.

We see those options in the following screenshot, along with the option to use Power Apps, which we will discuss in *Chapter 6*:

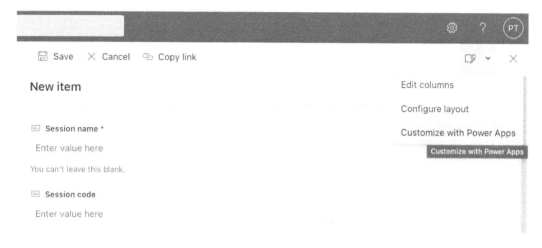

Figure 5.10 – Customizing the new item form

The **Configure layout** option opens a screen where we can paste the JSON formatting code. This provides a way to change the layout and order of the list forms as well as to have elements show and hide dynamically based on rule conditions. There is, unfortunately, no designer as we have with list views. This page provides a good introduction to the process and syntax: `https://docs.microsoft.com/en-us/sharepoint/dev/declarative-customization/list-form-configuration`.

While JSON formatting opens up a world of possibilities, it does require comfort with development concepts and the syntax of JSON as a language in general. For simpler forms, such as polls, quizzes, and surveys, Microsoft Forms may provide a more user-friendly alternative.

Using Forms for surveys and polls

Microsoft Forms is a separate app that is part of your M365 toolbox. If JSON formatting is the chef-prepared four-course meal requiring some skill that not everyone possesses, then Microsoft Forms may be the higher-end microwave meal. It still hits the spot but requires minimal effort on our part.

Forms provides a quick and easy way for users who are not especially technical to build data entry forms and view the data that results, both in a raw form and via some automatic data visualizations. We've probably all seen and completed a Survey Monkey survey on the web. This is essentially the Microsoft answer to that technology.

Given that surveys in classic SharePoint don't have a modern equivalent, **Microsoft Forms** provides a viable alternative. For simple data entry forms, this tool can also be considered a partial replacement for InfoPath. Let's take a look at creating a new form and the options we have.

Creating forms

All users who have Forms turned on as part of their M365 license can create a new form to share within and outside the organization. Forms may be created and completed in either a browser or as a tab inside a team. A form is created by adding a sequence of questions using one of the built-in data types. Some of these are like SharePoint data types, while others provide a bit of a richer interaction. The data types are as follows:

- `Choice`
- `Text`
- `Rating` (1-5 using either stars or numbers)
- `Date`
- `Ranking` (basically a choice field where the options can be ordered to indicate rank)
- `Likert` (a matrix of statements and value choices 1-5 that can be customized)
- `File upload`
- `Net promoter score` (a *1-10 how likely are you* score range)

These questions can be added into one or more sections. Initially, a single section is provided, but additional ones may be added. Questions and sections both support branching so can skip forward based on the answers that are provided.

When a form is ready to be distributed to potential respondents, we have a few options that can be found on the settings for the form accessed via the *gear* icon:

- **Share to collaborate**: Allows others to own the form with you as an editor or respondent
- **Share as a template**: Allows others to use your form as a template but excludes responses
- **Send and collect responses**: Sends the form to be completed

We can specify who our respondents are. We can allow anyone to respond, which is an anonymous survey. Anyone inside or outside the organization can respond to the survey, but we can't track responses against respondents. We can open the responses to everyone in our organization. This means that any licensed user could complete the survey. We have the option to capture the names of the respondents or allow anonymity. The last option is to provide a specific set of people in the organization to be respondents.

We can also set a start and end date for the survey. We can shuffle questions, show a progress bar, customize the post-response thank you message, and limit the responses to one per person. The responses for the survey can be seen by the creator or owners by clicking on the **Responses** tab. Each question is displayed with a visualization in keeping with the data type and is interactive. In the following figure, we see an example of the **Responses** page:

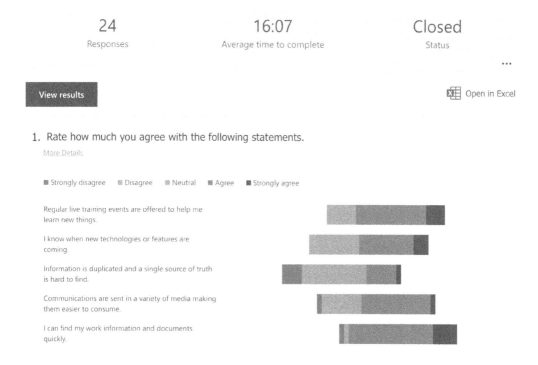

Figure 5.11 – A Likert scale question results

Rather than saving the data to a SharePoint list or a file that resides in SharePoint or OneDrive for Business, the data is saved in what is essentially a spreadsheet in the cloud. We can export and download the responses to an Excel file, but the visualizations are only available when viewing the responses on the web. A very common request I hear is to have the data from Forms written to a SharePoint list. While this isn't possible out of the box, we can use Power Automate to connect to both services allowing Forms to be your frontend and SharePoint to be your backend. We'll explore Power Automate in *Chapter 6*.

Microsoft Forms is a great option for non-technical users to create simple data entry forms and to get useful data visualizations without a tool like Power BI. The option to add a form as a Teams tab or to embed a form into a SharePoint page via a web part allows the form to be accessed from a variety of places. Next, let's turn our attention from Forms to media, and discuss the role of Stream in our collaboration toolbox.

SharePoint, Stream, and video files

Microsoft Stream is the video streaming service and has been a part of the M365 toolbox since 2017. Until recently, it was a standalone service that could partially integrate with M365 Groups. It replaced a deprecated (and short-lived) SharePoint video service. The idea was that we would have a central repository for video files within our tenant and a way to create and attend live events. That central repository was the Stream website, `web.microsoftstream.com`. Videos could be organized and searched in a few ways:

- By the groups that owned them (tied to M365 Groups)
- Into channels (a curated list of videos that could be followed)
- By people who posted them (search by someone's name)
- From meetings (recorded meetings from Microsoft Teams)

In addition to its own site, Stream also provides a web part that can be added to modern SharePoint pages.

The Stream service brought some intelligent features such as noise suppression, autogenerated captions, automatic transcription, and deep search capabilities that allowed us to perform a text search on the transcript. These video files were stored in their own dark corner of Azure Storage, however. This meant that the files were not subject to the same retention and compliance as other files in M365. It also meant that sharing a Stream video outside the organization was impossible without downloading a copy and sharing it from somewhere else.

To be fair, we're speaking of Stream in the past tense, but it's not entirely gone. Many of my customers have a substantial amount of content in what we are now calling Stream Classic, which will eventually be retired and replaced by Stream 2.0 or Stream on SharePoint. Let's be forward-facing and explore how Stream and SharePoint are going to be better together.

Stream on SharePoint

The transition is already taking place, but the future state is one in which audio and video files will be stored in either SharePoint Online or OneDrive for Business. So, we'll be able to work with videos from Teams and Yammer just like Word documents or PowerPoint slide decks.

Another immediate advantage is that these files will be brought in line with retention and compliance tools and can be shared easily with the organization or with guests just like any other file type. This also means that SharePoint once again becomes the scaffolding and backbone of another collaboration service in the cloud. To come back to our cooking metaphor, we can consider that Stream Classic was a single-use tool, like an avocado slicer or a garlic press. Rather than taking up space with limited tools, we can just use a knife to do both tasks and several more. SharePoint continues to be our multi-use tool.

While the transition is taking place, we will have video files that live in multiple places. We may have videos in the classic service that will need to be migrated. We may have new video files that are being uploaded directly to document libraries or through the new Stream application in preview at `stream.office.com`. This new page shows video files across sites, Teams, and OneDrive locations that we have access to. The following figure shows the UI of the **Stream on SharePoint** application:

Figure 5.12 – The new Stream application showing content from multiple sources

In the screen capture, we see videos sourced from OneDrive locations and a public team. We can mark videos as favorites or filter them via keyword search and sort by name, date opened, owner of the content, or last activity.

One of the most likely sources of video content in this new world of Stream is from Teams meetings. When using Stream on SharePoint, a standard (non-channel) Teams meeting that is recorded is saved to the OneDrive of the organizer if automatic recording is enabled, or to the location of the person who clicks the *record* button otherwise. If external users are in the meeting, they will not have immediate access, but the owner of the recording can share it readily from OneDrive. We can find the `.mp4` file in a folder called `Recordings`.

For channel meetings, the file is going to go to the `Documents/Recordings` path in the team channel that contains the meeting. The member who clicked *record* will have edit rights; every other member will automatically have access to the recording file as well.

Viewing videos

In addition to viewing a video on the new Stream application page or viewing a Teams meeting recording from the recording link, we also need to explore new options for viewing videos on SharePoint site pages. The Stream web part exclusively works with Stream Classic, so if our videos are stored in SharePoint, we need another option.

For single videos, we can use the **File Viewer** web part. This will play the video file on the page with the ability to pop out to a full-screen experience. To replicate the ability of the Classic Stream web part to connect to a channel and display thumbnails of multiple videos, we'll need to use the **Highlighted Content** web part.

For example, if a department has a monthly all-hands channel meeting in Teams that is recorded, we can add a Highlighted Content web part to the page and choose the filter option. Our content sources could be multiple sites, but a channel meeting is connected to a team that has a team site running behind it. We can select that team site and use our filter to only find videos with `all-hands` in the title. We'll sort by the most recent so we can view the last few recordings. Our layout will determine how video thumbnails and descriptions will display, and we can choose a certain number to include. A **See all** link will automatically be generated to view more. Other options and additional configuration details may be found here: `https://docs.microsoft.com/en-us/stream/streamnew/portals-set-of-videos`.

Islands in the Stream – making the transition

Given that we are in a period of transition for Stream, what flows on through to the other side of that transition and what stands in the middle blocking our way? Let's take a look at some features that no longer work, have been replaced, or provide new functionality.

What stays behind in Classic Stream (or at least isn't currently available in Stream on SharePoint)?

- Editable transcripts of recordings and manually adding a caption file to an existing video: A VTT file where you can tweak your transcript

- Embed codes for videos: The ability to use an embed element on web pages

- A rich video player with comments and transcription: such as videos on YouTube

- A dedicated mobile app: A downloadable Stream app for viewing video

- A tool to trim the beginning or end of a video: Without needing a separate tool

- The concept of video channels: A way to organize videos topically

What flows through or is new to Stream on SharePoint?

- Videos as a content type for content filtering: Filtering documents in a library and only choosing videos

- Organizing videos by folder, metadata, or enterprise keyword: The same capability for any other document

- A new player page that will restore searchable transcripts: A web part that restores some lost capability

- Jumping to a part of a meeting where a specific slide was presented or something was said: Not having to slowly scroll through but jumping straight to a point of interest

- Transcript editing in a dedicated pane: Editable transcripts are restored and have their own spot for the work to be done

- Playlists as an alternative to channels: A new and improved way to organize video content

A migration tool is currently in preview to help us move videos from one Stream to another. We'll be able to connect a source to one or more targets in SharePoint and OneDrive. The tool will include reporting along with migration assistance for content, organization, metadata, and permissions. Content will be scanned to produce a list of videos available for migration in one of five container types:

- Microsoft 365 groups (videos in a group)

- User containers (videos in MyContent)

- Stream group (videos in a classic Stream group)

- Company-wide channel (videos in a company-wide channel)

- Orphaned container (videos with no owner)

The Stream migration tool will be available from either the Stream or SharePoint admin center and will essentially work as a connector in the **Migration Manager tool**. The following figure shows the **Migrations** tab of the tool after scanning and mapping have already taken place:

Figure 5.13 – A screen capture of the Stream migration tool

CSV templates can also be used for bulk uploading to destination paths, as we discussed with SharePoint migrations in *Chapter 2*. Migration logs and a migration summary report will help us to verify the videos that have been migrated.

In this section, we've seen that Stream is changing the backend storage to one based on SharePoint. Video files will be like any other file, and enhancements to viewers, transcriptions, and sharing will be released over time. Next, let's look at one more service built on M365 Groups that helps us manage our tasks and work.

Making plans and managing tasks

There are different styles of cooking, different methods of preparation, and different appeals for varying meals. Yet, a common goal exists of providing nourishment and enjoyment whenever and however we cook. While streaming video is a very different type of service than meetings, document management or coordinating on tasks and work, they all function together to produce a highly collaborative and informed workforce. The tools each provide their own flavor to the overall dish.

Planner is a service that helps groups of people collaboratively complete projects and work on tasks on the web at `tasks.office.com` or on the Planner mobile app. Each plan is a kanban board containing buckets for organizing tasks. Each task contains data elements to indicate assignment, completion status, notes, checklists, and color-coding/tagging. The following figure is an example of a Planner plan in action. This view is the same whether we are on the Planner website or when a **Planner** tab is added to a Microsoft team:

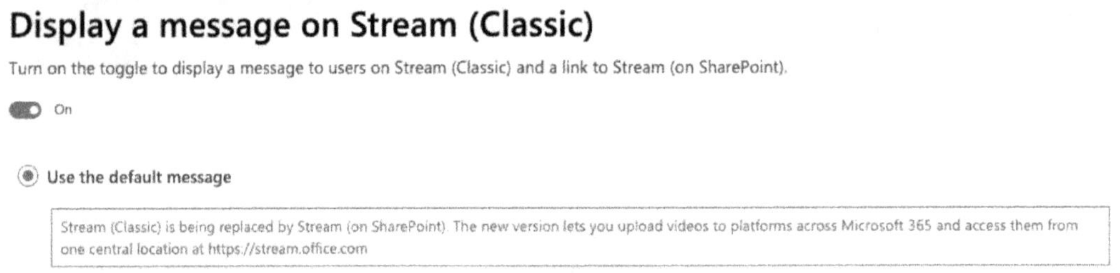

Display a message on Stream (Classic)

Turn on the toggle to display a message to users on Stream (Classic) and a link to Stream (on SharePoint).

🔘 On

⦿ Use the default message

> Stream (Classic) is being replaced by Stream (on SharePoint). The new version lets you upload videos to platforms across Microsoft 365 and access them from one central location at https://stream.office.com

Figure 5.14 – A Planner plan in board view grouped by bucket

In the figure, we see three buckets to organize tasks. All we must do is drag the task between buckets to reorganize them. Another organizing concept is the label, which can be defined by each group to provide metadata for their tasks. One or more people may be assigned to complete a task, and images on each card show the progress. The red background on dates lets us know that a task is overdue. Toward the upper left, we see that this is a Linked plan. That means that the plan is linked to other content, more specifically, it is a plan that is linked to a channel in Microsoft Teams.

The **Charts** view shows visualizations to let us know how we're doing on the status of tasks, how many tasks there are per bucket, how many have been completed, and how many tasks have high versus low priority. A **Members** view allows us to see whether a member is overloaded, underworked, habitually on time, or past due on completion. The **Schedule** view displays a calendar layout where task due dates are displayed. There is a panel there that also allows us to see unscheduled tasks.

Every M365 group gets a Planner plan, whether it's used or not. That means the people who can be assigned tasks are all members of the team. Since we also know that an Exchange Online calendar, email, and a SharePoint site are always provisioned for a group, we can be certain they are there to support the plan as well. These additional resources and functions can be found by clicking on the ellipses next to **Schedule**, as seen in the following figure:

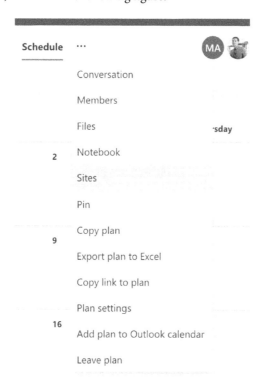

Figure 5.15 – Accessing Exchange and SharePoint resources for a Planner plan

The **Conversation** menu item opens the shared **Group** mailbox. Since the plan we've been viewing is connected to Teams, it may make more sense to have conversations take us into channel chat, but this is one of the few places that mailbox gets exposed. The **Files** tab, as may be expected, opens the **Documents** library in the associated team site. Although we do have a plan connected to a Teams channel, the library opens to an **All Documents** view at the top level, displaying all folders and files.

While we may simply go to the site to view any documents, there may be some that have a close relationship to individual tasks. Each task does have an **Attachments** area where files may be linked to the task. We can choose to have the document displayed on the card, which adds a live preview of the document. It also adds these attachments directly to the root of the library. This means that SharePoint stores the documents for us, but those files have no visible connection back to the plan tasks from the SharePoint side.

In the menu of the plan, there is also a link to **Sites**. This takes us to the home page of the team site. We can add a Planner web part there, configured to show the board, all charts, or a single chart. This allows the site to be the only UI we need to manage our plan if desired. Of course, other modern web parts may also be added so that we have elements such as the **Events** web part showing key milestone dates, a file viewer showing the project charter, and so on.

So, we see that Planner is an excellent tool built on M365 Groups and SharePoint for allowing a set of people to collaborate on tasks and content together. There is also another tool in the toolbox focused on helping us track our individual tasks.

Personal tasks

The tool called **ToDo** is the personal task management tool in M365. We have a web (to-do. office.com), desktop, and mobile device apps to interact with. Assuming we log in using an M365 work account, we can view tasks that are manually added in the tool but also get flagged emails from Outlook and planned tasks from Planner. In the following screenshot, we see a task that comes from the Mark 8 Project Plan:

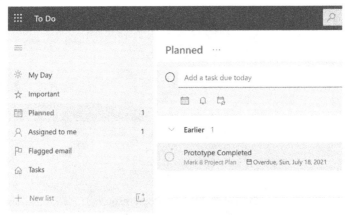

Figure 5.16 – Viewing planned tasks in ToDo

Clicking on the task allows us to mark it as complete in ToDo and have that update get passed back to Planner. We can add the task to the **My Day** view to help create a daily task planner, or we can open the task directly in Teams and interact with the Planner UI there.

ToDo is more of a personal application that can gather tasks from Planner. When we create tasks, we get the option to add files in a fashion similar to Planner. We may guess that those files would go into OneDrive for Business, as we have seen with tools such as Teams. However, Exchange Online is used in this instance.

The two task management worlds are so tightly linked together that they are a single app in Microsoft Teams. The app **Tasks by Planner** and **To Do** looks a great deal like the **ToDo** interface, but there is a Shared plans area where the Planner plan tasks that have been assigned to me reside. This is a great workaround if you need a single place to view all Planner tasks assigned to you. This is a notably missing feature for project and program managers who need to see this for any member they choose, but individuals can see their own.

Summary and planning document

In this chapter, we've seen that SharePoint is no longer a standalone platform. Site pages and document libraries provide direct and indirect support to anything M365 Groups-related, but play a special role in the successful implementation of Teams, Yammer, Lists, and Planner. The question is never whether we should use SharePoint or something else. The answer is almost always SharePoint *plus* something else. SharePoint is like salt and pepper. It's hard to cook without it and why on earth would you want to?

Here are some things to consider when preparing our planning document where SharePoint and related services are concerned.

Teams and communities

- For each need, should we create a community for org-wide communication or a team for tighter collaboration?

- Do we have a need to have both? Remember that an M365 group connected to a Microsoft team can't be used for a Yammer community and vice versa.

- Should the same people who have access to chats and discussions have access to the site and documents?

- Would we like to share a SharePoint site with certain people without giving them access to all the other Groups-related content?

- Do we need to create an all-company team or community for the entire organization?

Microsoft Lists

- Have lists been widely used in SharePoint in the past? If not, why not?

- Do people own spreadsheets or Access databases that could be converted to a list instead?

- Do people know they have the option to create private work-related lists in the Lists app using OneDrive for Business?

- What are some visualizations that would help with the adoption and consumption of lists?

- Are there list input forms or views that would benefit from JSON-based customizations?

Forms for surveys and polls

- Are there any other survey tools that are being used in the organization that may have a subscription cost associated with them?

- Which people and departments may have recurring needs to create simple forms, surveys, or polls?

- Are people aware that Forms can be used as messaging extensions in Microsoft Teams meetings to embed a poll directly in meeting chat?

Video files

- Is the organization currently using Stream for video?

- Which videos and channels would be the easiest or quickest to migrate?

- Do people in the organization know how to share video recordings?

- Which sites would benefit from linking to or embedding video content on their site pages?

- How important are editable transcripts and captions for your video assets?

Managing tasks

- How widespread is the adoption of Planner? It's hard to tell just looking at Groups usage because they all have associated plans.

- Are there other project management tools such as Trello in use that may be costing money that could be replaced by Planner?

- Are the project managers aware of Planner as a tool and its advantages?

- Do people know they can use the Tasks by Planner and To Do app in Teams for a unified task management experience?

In the next chapter, we'll continue to further explore how the toolset in M365 can help us leverage the power and capability built into SharePoint and to take it to the next level of functionality and productivity. We'll explore the tools within the Power Platform.

6

Making SharePoint More POWERful

Home improvement stores have done big business over the years thanks to our desire to do more and more work by ourselves. My wife is an interior designer, so I've resigned myself to treating this as something of a hobby. Many of us choose to do it ourselves, however, to avoid the time and expense it may take for a professional to do the work instead. The Power Platform from Microsoft can be a powerful way to address business process automation and low-code development needs by adopting the DIY, self-service mindset.

In this chapter, we'll explore how SharePoint can be augmented, extended, and strengthened by using Power Automate and Power Apps. We'll also explore Power BI in a limited way as a means of building richer user interfaces for list metadata. Considering how the Power Platform can connect to the entire toolbox of M365 and well beyond, we'll also discuss leveraging these tools outside of the SharePoint context as well by looking at these topics together:

- The low-code revolution – introduction to Power Platform

- Workflow implementation with Power Automate

- Forms customization with Power Apps

- Visualizing data with Power BI

- Using connectors to extend Power Platform across M365

- Power Platform governance

By the end of the chapter, we will have explored the tools within the Power Platform, what they're capable of, and how they are designed to integrate with SharePoint Online. These integrations will enable us to customize forms, create workflows, visualize data, and connect SharePoint to other tools and services.

The low-code revolution – introduction to Power Platform

Let's face it. Users of technology have always looked for ways to get their work done with the least difficulty and delay possible. When we meet barriers that slow us down, we naturally want to find creative ways around them. This is often referred to as *shadow IT* and can lead to a variety of unknown, unmanaged, and untrusted apps, sites, and tools. It's only natural that IT professionals will seek to protect the organization and themselves by restricting access and limiting features.

SharePoint has been no exception. When reporting on data was limited, we just connected Excel to a list. When we needed business logic in forms, we turned to InfoPath. When we wanted a more complete app experience, we built access databases for critical business processes and maybe even hosted them in SharePoint. People in our organizations have often become citizen developers, despite efforts to limit their toolset or contain their curiosity.

Microsoft's Power Platform is a suite of tools designed to help make our IT DIY dreams a reality. Let's start by looking at an overview of each tool, and then we'll dive deeper into how they work and how our architecture planning is impacted by each tool:

- **Power Automate** – Workflow and process automation tool, useful as a **SharePoint Designer (SPD)** replacement

- **Power Apps** – Low-code/no-code development tool, useful for customizing List forms and building standalone apps, especially for mobile

- **Power BI** – Reporting and data visualization tool, useful as **SQL Server Reporting Services (SSRS)** replacement or to enhance List views

- **Power Virtual Agents** – Chatbots largely accessed through Teams, useful as a Q&A replacement, but mostly out of scope for this chapter since these are not surfaced in SharePoint

These tools together are designed for IT professionals, developers, and everyone else in the organization to build the apps and automated processes that boost productivity and empower users. These tools also integrate with SharePoint Online. Across the top of any modern list or library, we'll find menu items related to the Power Platform. We can choose to initiate our integration with SharePoint data from here or start from the Power Platform apps themselves and create a data connection. In the first section, let's take a look at how to enable process automation around our SharePoint data with Power Automate.

Workflow implementation with Power Automate

For Power Automate, think workflows. Power Automate is the cloud replacement for SharePoint Designer workflows but is much more capable and extensible. While there are integrations with SharePoint Online, and that is where our primary focus will lie, Power Automate is its own M365 app that is capable of a variety of flows that connect and control cloud tools and processes as well as those on a device running Windows.

We get a glimpse of the capabilities by logging into the web app and creating a new flow. This shows the options for creating workflows in Power Automate, which we will explore next:

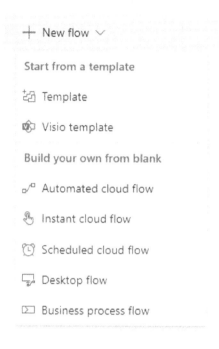

Figure 6.1 – The list of options when creating a new flow

First of all, we see five different types of flows that can be created:

Automated cloud flow – Similar to an SPD workflow that triggers automatically based on an action, such as creating a new list item or receiving a new email.

Instant cloud flow – Manually triggered workflows from the Power Automate site, via a hyperlink, or from a flow button in the Power Automate mobile app.

Scheduled cloud flow – Probably most like a timer job in SharePoint on-premises. We can run a set of actions at specific times on a recurring schedule.

Desktop flow – Otherwise known as **Robotic Process Automation** (**RPA**), these flows are created using a downloaded app and can work with a variety of scripting languages and local computer resources such as peripherals and files, and integrate with UI elements such as message boxes, windows, and form completion. We might think of these like macros, as a screen recorder can generate actions dynamically.

Business process flow – These flows originate from Dynamics 365 and create a visual canvas composed of up to 30 stages, each containing steps to guide users through a process to completion. Per-user licensing is required, as is the usage of Dataverse.

While the capabilities are vast and varied, automated cloud flows and instant cloud flows are the most likely to be used to provide process automation where SharePoint data is concerned. Scheduled flows are also very common when trying to reproduce long-running SPD workflows or timer jobs. With SharePoint on-premises, there was a direct 1:1 link between an SPD workflow and the list, library, or content type it was designed to integrate with. With Power Automate, SharePoint is merely one of the many available connectors.

SharePoint integrations

One common request I see repeatedly is using Power Automate to implement an approval process for things such as requests or document publishing. Let's walk through a scenario of adding that workflow to a SharePoint list.

We can start with the **Travel requests** list template as seen in the following figure. The list tracks requests for company travel, which would certainly require approval since planning and costs are involved. A column called **Approved?** is already part of the template, but is manually set to **yes** or **no**.

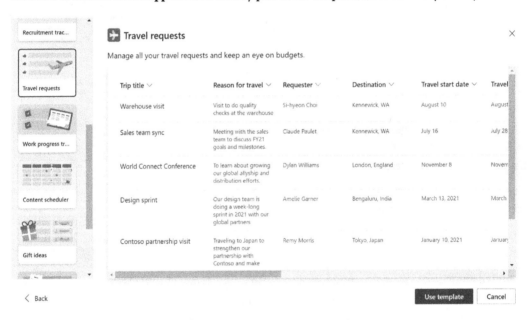

Figure 6.2 – The Travel requests template used to build a SharePoint list

To automate the approval process, we need to create a flow that will run whenever a new request item is added. Depending on the actions that we want to run when the item is added, we can either choose the **Automate** menu item or the **Integrate** menu item, both of which are seen in the following figure:

Figure 6.3 – The UI for the Automate and Integrate menus on modern SharePoint Online lists

Automate options change based on column data types present in the list and generally are for built-in tasks created by Microsoft or flows we create with selected triggers. For example, since we have start and end date columns in the template, a **Set a reminder** option exists for both start and end dates. Clicking this would automatically create a scheduled flow in Power Automate to send an email notification in advance of the dates, based on the number of days in advance that we provide.

The **Rules** menu option allows us to create simple sets of actions based on triggers. These are automated cloud flows that avoid the Power Automate canvas for those whose needs are simple or whose appetite to build flows is limited. A trigger for a rule could be based on the following:

- A column changes
- A column value changes
- A new item is created
- An item is deleted

Since our example is an approval process that starts when a new request is added, we could choose *option three*. This is just a simple notification workflow, so all we can provide is the email address of someone who should get notified of the item creation. This isn't dynamic or based on list data, but rather a hardcoded value. If that's all we need because only one person in the company ever approves travel requests, then we need to go no further. For more complex workflows, we need to turn to the **Integrate** menu option instead.

Under **Integrate**, we would choose **Power Automate | Create a flow**. This opens a panel on the right where we are prompted to choose a template. Chances are, we'll find one that fairly closely matches our needs or could be edited to do so. There is, in fact, a **Start approval when a new item is added** template as the second choice in the list. If we want to start from scratch with a blank canvas, we will need to visit the Power Automate app instead.

Each tool or system we integrate with requires a connector. Having chosen the workflow template noted previously, we now notice there are four connectors required for the flow to function (as seen in *Figure 6.4*). The connections are made using the credentials of the user creating the flow. Since this flow sends emails, retrieves SharePoint data, creates an approval within Power Automate, and gets user attribute data, the four connectors are necessary for that work to complete:

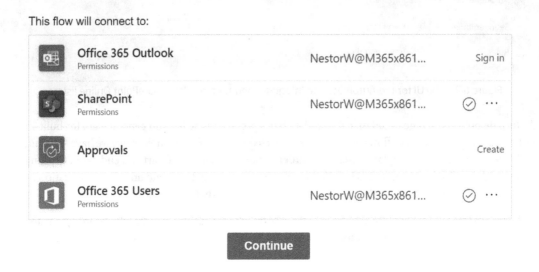

Figure 6.4 – The connectors required for a flow to gather and process data

Flows are a combination of a single trigger with multiple actions, and optional steps to group actions together. As we continue with the flow, the trigger is already set to look in our site at the list from which we initiated the flow. If we began from a blank canvas, we would have to choose the SharePoint connector and manually supply that information. The first action is the approval action, which starts upon the creation of the item. This will result in an entry into the **Approvals** section of the Power Automate web app and an email being sent to the person the task is assigned to. That email will contain an actionable link that will allow the recipient to approve or reject the request without having to go to SharePoint or the Approvals Center at all.

In the action, we can specify the approver by entering a name or email directly or by using dynamic content that allows us to get data from the SharePoint list item or another connected source (shown in the following figure):

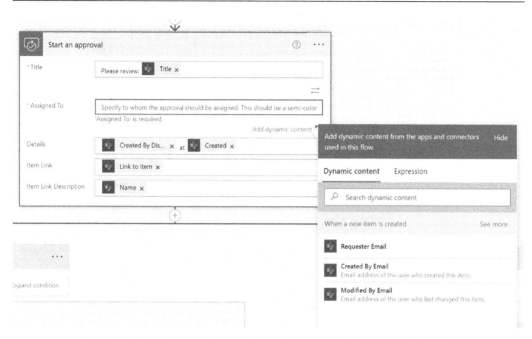

Figure 6.5 – Dynamic lookup for the Assigned To field in the approval workflow

There are three people fields in the SharePoint list so we could choose any one of those. Since there is no approver column in the list, we could go back and add one. Upon refreshing the workflow, we could then choose it in the action. We could also send the request to the manager of the requestor. There is an action in the template to get the user profile of the person who made the request. Since that action is introduced after the approval action, it doesn't provide that data in the dynamic lookup. This is where altering the flow created from the template would make sense.

We may also make other changes. We could (and should) replace the **Start an approval** action with **Start and wait for an approval**, which would allow us to have multiple approvers and decide whether we need all of them or only one of them to approve or reject to move things forward. **Start an approval** is also deprecated and only available if we start with a template. This action would also allow us to create our own values for custom responses, in case the words *approve* and *reject* don't capture our requirements. To add actions, we can click on the plus icon on the canvas and choose an operation, as seen in the following figure.

By filtering with the word `approval`, we can avoid searching through hundreds of available actions:

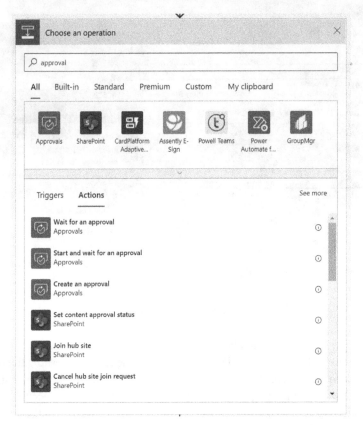

Figure 6.6 – Adding a new action to a workflow, filtered by a search term

Once the workflow is completed and saved, we can test it from the design canvas or by adding new data to the list. We can view a history of the flow's execution as well by leaving our design view and looking at our cloud flows. Each run of the flow over the last 28 days can be clicked on to see the steps that were taken, data that was exchanged, duration of each step, and overall status.

Once our flow is tested, a good governance idea is to share the flow with someone else as a co-owner. This could be one or more individuals or user groups. Once that sharing has happened, others can make edits, if necessary, when the initial creator can't be reached or if they leave the organization. A note to remember that may save you some time and aggravation – when shared, the flow no longer appears under **Cloud flows**, but instead **Shared with me**, which is the case even if you are the one that did the sharing.

In this section, we've seen that the Power Automate tool allows us to configure automated processes to run as we interact with data in SharePoint, replacing tools such as SPD. Another common need is to customize the look and feel of the form associated with adding and editing items to lists. For that, we need Power Apps.

Forms customization with Power Apps

Power Apps is a tool that can be used to customize the UI for SharePoint list forms, in a similar way that first-party tools such as InfoPath, or third-party tools such as Nintex Forms may have done with classic SharePoint. It's important to keep in mind that Power Apps is not just a way to extend SharePoint, but so much more as well. Power Apps is essentially a low-code developer tool to create the frontend of an application that can run through the web, in Microsoft Teams, or on mobile devices. Power Apps supports connectors in the same way that Power Automate does, so it can connect to SharePoint along with hundreds of other sources, tools, and apps.

Power Apps can be used to create the following:

- **Canvas apps** – Apps built by dragging controls onto a design canvas and connecting those controls to data through connectors. This is what gets created when we customize SharePoint list forms. Canvas apps support PowerFx formulas.

- **Model-driven apps** – Suitable for more complex application scenarios, but data-first, based on underlying data defined in Dataverse.

- **Portals** – A website where both internal and external users have secure access to your data in Dataverse. Identities don't need to be tied to Azure AD. Power Apps portals support an anonymous license model, but if the user logs in, their identities are tied to Azure AD either as an internal user or external B2B user.

So, Power Apps really is a developer's tool that doesn't require all the developer's knowledge. Let's focus on how Power Apps and SharePoint lists work together to create a customized UI for new and edited item forms.

Power Apps and SharePoint

Similar to Power Automate, the integration between SharePoint and Power Apps is built into the menus of modern lists and libraries. It is also a standalone site, desktop app, or mobile app that can start with a blank canvas and be connected to SharePoint data. If we continue with our customization of the *Travel requests* list built from a template, we can change the look and feel of the forms that open in the right-hand panel by choosing the **Integrate** menu and selecting **Power Apps**.

We have the option to create an app or see all apps, but the real magic is with the menu option to customize forms. We can find the same link by creating a new item or editing an item and selecting the icon in the upper right. Here, we have the option to edit columns or configure the layout, which would not involve Power Apps, or to use **Customize with Power Apps** (shown in the following figure):

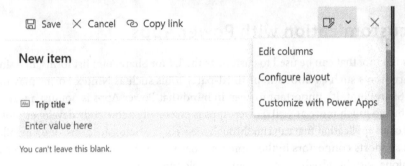

Figure 6.7 – Viewing customization options from a new list item form

Upon clicking the button to customize, we'll be taken to a new Power Apps screen and a canvas app will open. With custom forms, we have a `SharePoint Integration` object in the tree view. In canvas apps, the connection to the SharePoint list is there from the beginning, and controls for our list columns have been added. If we select **SharePointForm1** in the tree view on the left, we'll see the form in its default state. On the right, the **Data source** property has been set to our list. This connection is defined on the left side, under **Data**. Additional lists or other connections can be added here as well.

Below the data source, the list of fields may be edited along with the type of control. In the following screenshot, we can see that the **Title** field, which is text, can be formatted for a phone number, email, multiline, or even a dropdown of allowed values, all while keeping the column type the same on the list. We can also reorder the elements on the form, and change the fonts, sizes, visibility, and colors:

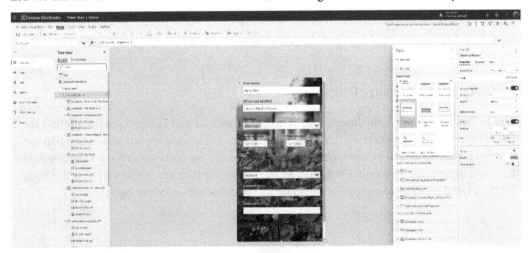

Figure 6.8 – Editing a list item form in Power Apps with fields selected and customizations applied

When our customizations are complete, we can click the *play* icon to preview the form in Power Apps or save it from the **File** menu. Clicking **Back to SharePoint** in the upper left allows us to publish to SharePoint to enable the new form to replace the default. Our Power Apps app now displays when we edit an item or when we add a new one.

A connected app

If we want to provide users with an app experience seemingly separated from SharePoint, we can also build a new app from a blank canvas app template. When taking this route, we must choose either tablet or phone layout, as canvas apps are not responsive by default. The primary UI elements in our canvas app (or the integrated form) will be galleries, forms, and cards, each of which may be connected to our list data:

- **Galleries** – A gallery is a list of records or items built from one of the built-in gallery templates. Layouts can be changed after creation and we can change the sorting, filtering, and highlighting options. What's possible with layouts is depicted in the following screenshot, including the image, title, and subtitle option:

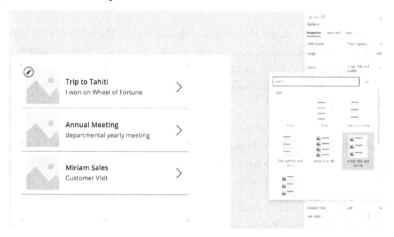

Figure 6.9 – Changing the layout of a vertical gallery control

- **Forms** – Whereas a gallery displays a list of items, a form displays a specific item to either view or edit. We can add a form to the app through the tree view by first adding a new screen to the app. Adding an item from the **Insert** pane will allow us to add a **Display** form from the **Input** category, which may then be connected to data. In this case, that's our **Travel requests** list. We can edit the connected fields we want to display.

- **Cards** – A form is comprised of multiple cards, each connected to a data element. In our example, we have a card for each list column that we've selected to display. In the fields list, we can change the control to render the data. If we need to customize further, we can unlock each card and interact with the **Advanced** tab for that card. This also enables the **Display** mode, which can be changed from **View** to **Edit** or **Disabled**.

To add our SharePoint data, we can select an object (such as our gallery) and choose the data source dropdown. Under **Connectors**, we should find **SharePoint**, where we can select the site and list that we'll use to build out the app. Recent sites may display the site we need, or it may become necessary to enter the URL. The lists will automatically populate after the site is connected. We can also add the data initially before building any screens from the **Data** section on the left of the canvas designer.

The gallery may be the place we start interacting with the app. To interact with each list item on its own, the next arrow icon to the right of each item in the gallery could be configured to open our form, which displays the entire item. We could add some *Power Fx* code to the `OnSelect` property of the icon, something like this: `Navigate(Screen2)`, where `Screen2` contains our form control.

We would continue down this path to build out the forms that allow us to fully interact with our data. We'll also need to build out the navigation and interaction elements as well. For example, even though the `Navigate` function mentioned previously would take us to the display form, we'll need to make sure we provide a way to get back to the home gallery. We'll also need to add buttons to submit data back to SharePoint if we make changes.

Generated apps

Taking the approach we've just walked through together will provide you with the most granular control over building your canvas app. It is, however, not the fastest way to build a new app. We can take a more data-first approach. While model-driven apps are usually defined in this way, we are going to stay in the canvas apps world but allow Power Apps to build out the basic UI from the data we have in our list.

In Power Apps, we can select **SharePoint** under the **Start with your data** section. We can choose our connection to SharePoint and our **Travel requests** list. Note that the layout is for phones by default. We can change this to tablets instead to have a larger design surface. The result of this effort is what's commonly referred to as a **three-screen app**.

We'll get a browse screen with a gallery control, a detail screen for viewing item information, and an edit screen for making changes or entering new items. Controls for sorting, refreshing, adding data, searching data, and navigation will be created and supplied with the necessary Power Fx code to make the app functional. All the forms and cards will be connected to data, though we can make whatever changes we would like. In the following figure, we can see the default state of the three screens side by side.

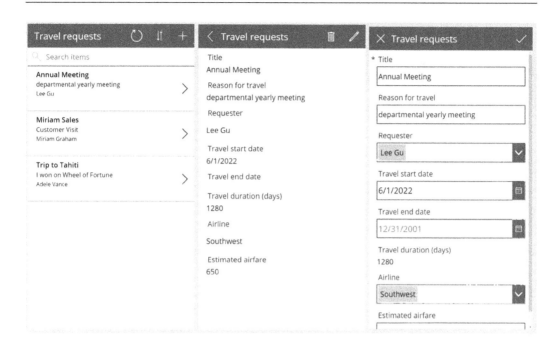

Figure 6.10 – Screenshots of a three-screen app generated from a SharePoint list

Generated apps give us a very rapid way to present and interact with the connected list data from SharePoint. Generated apps or custom canvas apps need to be deployed in order to be used.

Deployment

When we customize SharePoint forms with Power Apps, the changes we make are directly connected back to the list, so the deployment strategy is simple – just change, save, and publish. If we create our own apps from scratch or from data, we need to make the app available using one of the available deployment options.

First, we should make sure that the app is saved to the cloud so we can open and make changes to it from the Power Apps web app wherever we are. Once saved, we can then share the app. We can share the app with others as co-owners or users. Co-owners can make changes with us. To make the app available for everyone in the organization, we can choose **Everyone** when sharing. We also need to verify that everyone also has the necessary access to the underlying SharePoint list.

So now, how will users get to the app we've created and shared? The following figure gives us some hints:

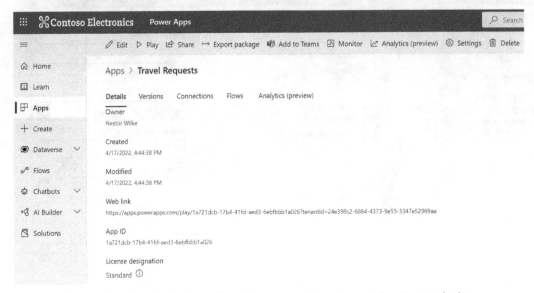

Figure 6.11 – The details for our Power Apps app with settings and options to deploy

Let's take a look at the options we have to get our apps into the hands of our users:

- **As a web app via a link** – A web link will be provided (depicted in the preceding figure), which we can use like any other URL. Maybe we can add a hyperlink to the site navigation so that users open the web app directly instead of going directly to the SharePoint list.

- **From the Power Apps site** – We can see any apps that we've created or that have been shared with us when we go to the Power Apps app from the app launcher in M365. Clicking on the app will give us the same experience as clicking the web link option we just noted. If we want to further the awareness and adoption of Power Apps, sending users to a listing in a place where they can create apps of their own may be preferable.

- **From Microsoft Teams** – The **Add to Teams** option is visible across the top of the app details page. Clicking here allows us to configure an app manifest, a JSON deployment file with details about our app, enabling deployment to Teams. This can happen by downloading the app package and uploading it as a custom app in the Teams app or admin center. We can also deploy directly to a team by clicking the **Add to Teams** button in Power Apps. We can then choose to add the app to a team as a tab in a channel or a private chat. The following figure shows us the deployment screen to add the app to Teams, either to a team or to a 1:n chat:

Figure 6.12 – Deploying a Power Apps app to Teams

The app then loads directly inside of Teams when we click on the tab.

- **From Power Apps Desktop** – For users on Windows, Power Apps Desktop can be a way to view and interact with Power Apps or to create them. When we authenticate with our M365 credentials, we'll see the same list here as on the Power Apps site. We can click to download any of the apps on the list. Once downloaded, the app can be pinned to the Windows start menu just like any other locally installed app or progressive web app.

- **From a mobile device** – We can interact with the Power Apps app from a mobile device by opening the link on a mobile browser or by opening the Power Apps app from Microsoft. Similar to the Desktop app, we'll see a list of available Power Apps apps. We can open the app from the list or add it to **Favorites** for quicker access. We can also create a shortcut or pinned app to open the app directly on our mobile device. On iOS, that takes the form of a shortcut, either by tapping or invoking Siri. On Android, we can pin the app icon to the home screen.

For mobile app deployment as described previously, we must use the Power Apps app to either open our canvas apps or to deploy them by using Power Apps to create a shortcut. As of April 2022, additional capabilities are in preview that will allow Power Apps to become a **Mobile Application Development Platform** (**MADP**). A feature named `Wrap (preview) for Power Apps` is not installed in tenants by default but can be added to Power Apps in the admin. Once installed, we can create canvas apps (or combine several) and then wrap them as custom native mobile apps complete with our branding and enterprise governance with Microsoft Intune. More information can be found here: `https://docs.microsoft.com/en-us/power-apps/maker/common/wrap/overview`.

This section has hopefully helped you to realize the capabilities of Power Apps to extend the UI associated with SharePoint list interactions, but so much more as well. We can connect to SharePoint along with hundreds of other sources to create our own apps to be consumed in a variety of ways. These apps give SharePoint a whole new dimension as a data source to be surfaced and consumed in heavily customized UIs, all while using a low-code or no-code approach. Next, let's take a look at one more option for presenting list data, but this time, as a report or dashboard, giving us new ways to view and summarize all the data in our lists.

Visualizing data with Power BI

Power BI is a Power Platform tool designed to create interactive data visualizations in the form of reports or dashboards. As with the other tools we've seen, Power BI is a tool available as its own app in M365, usable from the web, or as a downloadable Windows or mobile app. Power BI is licensed differently from the rest of the Power Platform. Power BI Pro is only included with M365 E5 licenses, not all licenses like the rest of the Power Platform. The app allows us to both create and consume content. In SharePoint on-premises, tools such as SQL Server Reporting Services were commonly used to create reports from SharePoint list data. Those reports no longer work with SharePoint Online, but their outputs can be replaced with Power BI.

While Power BI is a vast subject certainly deserving of a book all its own, let's focus on how Power BI can be used to quickly visualize list data in SharePoint. We can begin with the built-in integration within the **Integrate** menu on our **Travel requests** list (as seen in the following figure):

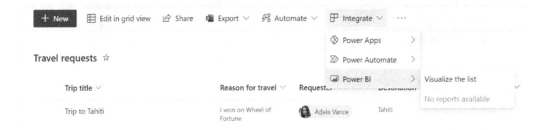

Figure 6.13 – Creating a SharePoint-connected report in Power BI from the Integrate menu

Clicking on **Visualize the list** will open a quick summary report (if you have an E5 license). This provides us with instant visualizations and insights that would not be possible just by using SharePoint list views, as powerful as they may be. For **Travel requests**, we automatically get summaries of hotel costs by month and year, the same for airfare, and a count of destinations for each time period as well. Each visualization can be personalized by changing the type of chart, the axes, and the values displayed.

The **Your data** section of the report is automatically provided using the list columns. Clicking additional columns creates new visualizations as part of the report. The **Filters** section allows us to further refine the data. The following figure is an example report that has been modified to show the data table and some additional reports on who is making requests, along with how many are made each month:

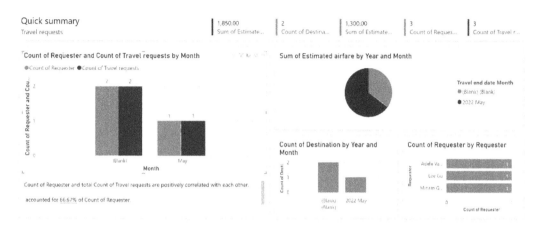

Figure 6.15 – An example Power BI report from SharePoint list data

In addition to what gets generated from the selected data, we can also choose the **Edit** button in the upper left of the report. This is part of what we see in *Figure 6.15*. When we switch to edit, we will have a much more elaborate design surface where we can add objects and visualizations, change colors and other visual details, remove unwanted visualizations, and add more data filtering options based on selected fields. The visualizations should be based on business requirements and provide the insights that are most pertinent. Multiple reports per list can be generated so we can choose to create several reports with fewer visualizations rather than one large dashboard.

When we complete our changes to the report, we can publish it back to SharePoint. The report is then visible as an option under the **Integrate** menu and opens in a new browser window in Power BI. Of course, we can always build new reports from scratch and connect to SharePoint data in the same way we've discussed with Power Automate and Power Apps, but the integrations out of the box with SharePoint allow us to create useful and insightful reports with minimal effort.

Next, let's briefly discuss the role of connectors to allow us to work with data across M365 and beyond.

Using connectors to extend Power Platform across M365

We've seen across the Power Platform tools that there is no intrinsic connection to SharePoint or any other tool or platform. This means that being able to connect to data, and to more than one source of data, makes these tools valuable for purposes well beyond SharePoint usage scenarios. We discussed connectors previously in the context of *Groups*, and the concepts are still the same. A connector allows us to interact with online tools and services in M365 and with other web-based tools such as Twitter, Trello, Bing, and Asana.

The AppSource site at `https://appsource.microsoft.com` gives us the most up-to-date information about apps built with connectors. If we filter AppSource on the three Power Platform tools highlighted in this chapter, there are 1,490 results at the time of this writing. We also have a listing of connectors available here: `https://docs.microsoft.com/en-us/connectors/connector-reference`. Given that we can include multiple connectors in a single app, flow, or report, we have the option to truly build applications that could be enterprising in scale and benefit.

For example, we could have a Power Apps app that gets drop-down list values from a local SQL server instance but gets written back as a choice option or text to a SharePoint list. We could build a form to gather employee satisfaction survey results and visualize those in a Power BI report. We could build a flow that could trigger whenever a task in Planner is created and send a notification to a Teams channel. We can leverage templates such as the *COVID-19 US Tracking Report* template app (`https://app.powerbi.com/groups/me/getapps/services/pbi-contentpacks.covid19ms`), which connects to *USA Facts* as a data source. The possibilities are endless.

Power Platform governance

With so many possibilities and so many ways to deliver value quickly to the organization, it is only natural that Power Platform adoption increases as those in your organization become more aware of its value and ease of use. We'll begin to see both champions and citizen developers. Champions will promote the business value of self-service tools that can make enterprise connections without the need for a developer or IT involvement. Citizen developers will be a community of makers that embrace the tools and create utilities for themselves and their departments that may become embraced across the enterprise.

Before that adoption and usage begin to take hold and grow, it is ideal to set some rules and best practices for the tools in the platform. We need to make some decisions on how the tools are to be used, which of those hundreds of connectors can be leveraged, and where apps, flows, and reports should be built. This is Power Platform governance.

Microsoft's Power CAT Adoption Maturity Model is a great guide for assessing and driving Power Platform adoption. It defines five stages of maturity, levels 100–500 at `https://powerapps.microsoft.com/en-us/blog/power-cat-adoption-maturity-model-repeatable-patterns-for-successful-power-platform-adoption`:

- **Initial** – Organic growth, chaotic implementation, ad hoc usage, and focused on individuals
- **Repeatable** – Documented processes, some central control, gathering metrics, and some policies
- **Defined** – Emerging standards, consistent environments, **Data loss prevention** (**DLP**), and process automation
- **Capable** – Defined metrics and monitoring, enterprise-scale, and established champions
- **Efficient** – Experts community, AI and other advanced features, **Application Lifecycle Management** (**ALM**), and promote ideas

An organization may fall into one of these stages as a whole, or different groups within the organization may be more or less mature in their usage of Power Platform. I think one of the primary outcomes of this model is to identify how effective and formal our governance is. Within the five phases, we often see organizations taking one of three common approaches to adoption and proficiency:

- **Minimal governance** – The following characteristics are usually true:

  - Citizen developers train themselves and begin to fill immediate needs.

  - Default licensing for Power Platform allows everyone access to the tools.

  - Apps are more local and less enterprising in scope.

  - All artifacts are created within the default environment.

  - Citizen developers have not formed a community of practice.

  - Very little formal documentation or strategy.

- **Standard governance** – The following characteristics are usually true:

  - Formal training for citizen developers.

  - IT sets standards for DLP and connector usage.

  - Multiple environments are actively used (usually dev, test, and prod).

  - The size of team driving app needs is growing in size.

  - A community of practice or Center of Excellence has started.

- **Strict governance** – The following characteristics are usually true:

  - More IT-focused with tighter licensing.

  - Citizen developers request apps that IT or CoP approves.

  - The number of citizen developers is smaller and more official.

  - Restricted access to lower environments.

  - The process from requirements to the app is more restrictive.

Within these descriptions, there are some key elements that factor heavily in the level of maturity and the scope of governance. One of the first concepts to be mastered is environments.

Environments

In its default state, Power Platform licenses in M365 allow any user to create what they need in a self-service way. Power Apps and Power Automate flows that we have spoken about so far have been created in the default environment. An environment is an area for hosting Power Platform artifacts. While all our environments are part of the same tenant, they do create a boundary for security and for limiting connections to data sources.

Each tenant gets a single default environment, created automatically when the first Azure AD user is created, where all licensed users create apps and flows by default. When we use the SharePoint list integrations, the default environment is the location where the items are built. If you have a license to Power Apps, you have access to the default environment in the **Maker role**. The Maker role means you can create and distribute resources within an environment. The other role is that of an **admin** who can define permissions and policies as well.

Automatic access to the default environment allows everyone to create what they need, though admin access is still granted explicitly. The default environment can't be deleted, nor can we remove the Maker role from users who are licensed there. The default environment can have only one Dataverse database, also true for other environments.

Additional environments may be created from the Power Platform admin center. These could serve the purpose of traditional application life cycle management where we have a development or sandbox space, a testing space, and one or more production spaces. We might also create additional environments as a means of allowing only certain permissions for users to be in the Maker role.

Though our intents for the environment may vary, the types are either **Sandbox**, **Trial**, or **Production**. Users with a Developer Plan license can create a Developer type as well. Additionally, each team in Microsoft Teams automatically gets an environment when the first app or bot is created, up to a limit of five environments plus one additional environment for every 20 eligible Microsoft 365 user licenses in your tenant. It can then be the target of the app or flow deployment and a more limited Dataverse for Teams database. Some limitations in Teams environments include lack of API access, permissions limited to team owners and members, and a significantly smaller database size (2 GB versus 4 TB or more for Dataverse).

Environments are also tied to a geographic location, so multiple environments may be needed in multi-geo tenants. The creation of additional environments can be restricted to only global or Power Platform admins, who also control the assigned users. Each environment should also be seen as an opportunity to implement good governance by limiting which connectors are available.

Environment DLP

DLP policies in environments are really data policies that configure or restrict which connectors are available and for what purpose in each environment. Business and non-business classifications create definitions around what connectors can be used together in a Power Apps app or flow. When we define data policies, we can also choose to limit the available connectors by blocking them. This can only be done for blockable connectors. Core ones such as SharePoint, Approvals, Dataverse, and other M365 tools and services can't be blocked.

For other third-party connectors, we can decide which ones to block by selecting the connector and blocking it entirely or by configuring the connector to decide which actions are allowed. For example, we can select the SalesForce connector and block it altogether, which would prevent it from being shown at all to makers in the given community, or we can configure it and choose from about 16 actions to mark as allowed. This is shown in the following figure:

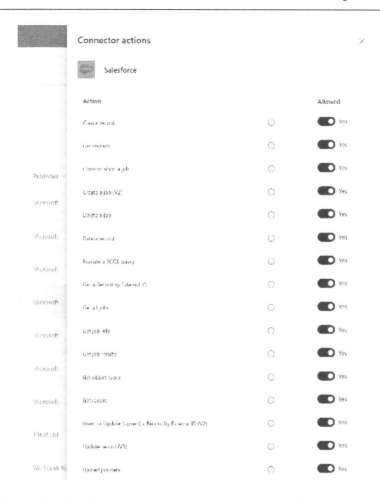

Figure 6.16 – Configuring connector actions in the Power Platform admin center

Being able to block connectors may also help to reduce inadvertent costs for using outside services in the Power Platform. Some connectors are standard while others are premium, which may not only have associated subscription costs but may also involve additional licensing costs in M365. Choosing to block these in a data policy for an environment makes sense. The **Class** column shows whether a connector falls into one or the other category.

In the configuration of a data policy, we can also choose which data group a connector is part of, *Business* or *Non-business*. Each connector begins life in the environment in the latter state, though that can be changed or set to blocked by default to limit the accidental impact. Making this distinction allows us to group connectors together based on whether they are more personal in nature or more suited for an organization's business needs. No business data should be allowed in personal apps or flows. With connectors grouped, they travel and work together, so a connector in the business data group can only work with other connectors in that same group.

Once defined, policies can be applied to one or multiple environments. We can also choose to exclude certain environments from a data policy. Environment creation should eventually become automated with approval flows that can be reviewed with environments created upon approval, rather than having to return to the admin center each time a new environment is requested.

The environment used by a specific app or flow can be changed. Apps, flows, and bots are stored when created in a specific environment. In the top-right corner, the environment currently used should be displayed. When clicked, a drop-down list of other environments available to a user is presented. Changing the environment will show the artifacts created in that environment. To switch environments, we need to export an app solution from one and import it into another.

This may be a better scenario for tasks such as ALM. Perhaps we've created an artifact in a sandbox and now want to move it to testing and then on to production. Clicking **Export** allows us to package things up in a `.zip` file that can then be uploaded in the new environment or even to a different tenant. We just need to make sure the credentials used for the connectors are still the same in the new spot.

Center of Excellence

To assist with the adoption and governance aspects of a successful implementation of Power Platform, Microsoft provides a free set of tools and components called the **Center of Excellence (CoE) Starter Kit**. The kit may be downloaded from here: `https://aka.ms/CoeStarterKitDownload`.

The kit is built on a Dataverse database (which means you'll need a Premium license) and provides automation and tools that help with monitoring, governance, reporting, and inventory. Not all components are required, but the Starter Kit is designed to assist with all aspects of Power Platform adoption. The core components first help you to establish an inventory of all apps and flows that makers may have already created. Admin components help with ongoing tasks such as DLP editing and permission assignment.

For the most effective results, these tools assist admins and makers to follow good governance practices, but the community of makers must also be able to support and inform one another. A Maker community should be established as well in Yammer or Teams to allow information to be communicated from admins and for makers to collaborate with one another.

Let's take a brief look at each of the components to understand the value each provides:

- **Core components**:

  - **Inventory** – What do we already have? All flows and apps are synced into tables in Dataverse and admin functions are built on top of the inventory data. The DLP editor and **Set App Permissions** are enabled as well. Multiple admin flows support the inventory gathering process. A daily sync keeps the inventory up to date.

  - **Environment request management** – Who needs to create what? This tool automates the request and approval process for environments and connectors. This app should be open to all makers.

- **Governance components**: Are we following the rules? These components work with the app auditing process to verify that apps with more than 20 users or frequently used chatbots have a business justification provided, that apps have been published within the last 60 days, and other thresholds are met. A traffic light indicates the level of compliance for each app. Business process flows provide a visual guide through the approval process but do require a Premium Power Apps license. Archival and cleanup tools are also available.

- **Nurture components** – Are we sharing best practices and examples? These components are all about the knowledge and reinforcement aspects of adoption. Training content as well as the ability to share templates and best practices for admins, makers, and users are defined here. The app can help coordinate events such as *App in a Day* and facilitate the desires of a maker to build a new solution:

  - **Maker assessment app** – Why are you building an app? We can ask a series of questions about the data, people, and support for a potential solution including licensing considerations.

  - **Template catalog** – This canvas app creates a repository for reusable templates that makers might generate. Templates will help us to enforce architecture best practices and reduce overall development time.

- **Theming components** – Does the app reflect our branding? A theme is built from a set of styles that define color, font, and other visual attributes for controls and components. A theme can be applied to the app that will automatically inherit the properties. A *theme editor* app may be used to create new ones and a theme gallery allows us to share our designs across the maker community.

- **Innovation Backlog components** – How do we develop ideas? The Innovation Backlog component helps us to gather and manage ideas and to estimate ROI. Makers may submit their ideas or vote on others, while also explaining the people, effort, and business impact of their proposals. This component can help us facilitate well-considered and designed apps before any type of development activities begin.

One additional and crucial part of the solution is the CoE dashboard. Here, we ask what metrics can we measure and what insights can we gain? The dashboard is built on Power BI and shows adoption insights such as how many flows, bots, makers, apps, and environments we have. We see usage patterns such as how many items were created in the last month and how many apps may be at risk of oversharing or being abandoned. We can also identify our top makers and which apps have the greatest usage and adoption. The dashboard delivers data related to each of the three primary areas of the CoE – monitor, govern, and nurture.

A CoE is all about the people who are making and using apps, implementing and following best practices, and sharing knowledge and information with each other. We can develop these concepts and processes on our own, but the CoE Starter Kit is a great way to implement a standard from which we can add our own flavor or customization. It's a highly recommended tool for driving any governance and adoption effort for Power Platform going forward.

Summary and planning document

In this chapter, we've looked at several ways that the Power Platform provides a replacement for on-premises SharePoint functionality. It is also a much broader and more potent way to enhance, automate, and visualize data and processes from across the M365 toolbox.

Power Automate provides workflow and process automation capabilities. Power Apps provides a way to build no-code/low-code solutions complete with business logic and data interactions. Power BI provides reporting and dashboard capabilities to visualize data quickly but robustly. Connectors provide the underlying capability that each tool relies upon to connect to M365 and third-party systems to provide either personal productivity or enterprise-scale applications.

The Power Platform is a set of tools on its own but integrates thoroughly with SharePoint, either from within the SharePoint Online UI or through the tools directly in their own apps and sites. When we plan for Power Platform integration with SharePoint, here are some questions to ask and details to consider:

Power Automate:

- Are there any SharePoint Designer workflows that need to be rewritten using Power Automate?
- What are some likely approval workflows that could be implemented?
- What are the most common data connections for workflows both within and outside M365?

Power Apps:

- Which lists would benefit from form customization with Power Apps?
- Do users in the organization consume corporate data through mobile devices?
- Who are the most likely citizen developers from both a business and IT perspective?

Power BI:

- Which lists provide business-critical data that would benefit from greater reporting functionality?
- Are there legacy SSRS reports that could be refactored as Power BI reports?
- Do report designers have knowledge of the underlying SharePoint list data structure?

Governance:

- Is everyone in the organization allowed to make flows and apps or should there be restrictions?
- Have citizen developers or makers been formally identified?
- Have Power Platform champions been identified?

- Has a Maker community been created in Yammer or Teams?

- Who will be the coordinator for the Maker community?

- Does IT control the creation of additional environments?

- Has the organization launched a community of excellence?

- Is training material available for potential makers to learn more?

In the next chapter, we will introduce a new section that will help us to understand and plan for the SharePoint site and compliance architecture. We'll start by discussing a feature called **hub sites**, only available in SharePoint Online, to group multiple disparate sites together in a flexible way.

Part 3:
From Tall to Flat –
SPO Information
Architecture

Part 3 provides a comprehensive and essential view into site architecture, all the varying types of metadata and labeling, and a security primer for SPO data.

The following chapters are included in this part:

- *Chapter 7, Up with Hubs, Down with Subs*
- *Chapter 8, The Mega World of Metadata*
- *Chapter 9, Keeping Things Secure – Permissions, Sharing, and DLP*

Up with Hubs, Down with Subs – Planning Hub Sites

Families are important. Functional families are indispensable. This is not only true for you and me but for SharePoint sites as well. While it's possible for all your organization's pages and documents owned by all your users in each of their business units to live on a single site, it's not recommendable. Too much family under one roof will end in trouble eventually. In SharePoint, each site can be a part of a greater whole but maintain its identity, permissions, and boundaries. How those sites work together has changed over time. We're moving away from tall, formal hierarchies and moving toward voluntary associations that can change over time to fit changing organizational needs.

In this chapter, we'll explore how to turn a site into a hub and how site families can work together to share information in consistent and compelling ways. We'll cover the following topics together:

- Site templates and the role of hubs
- Subpar subsites – the benefits of a flattening architecture
- Hub site benefits – roll up and push down
- Hubs and global navigation
- Viva Connections – bringing the intranet to Teams

By the end of the chapter, we'll see the value of organizing sites into hubs and embracing the flat architecture of SharePoint Online versus the subsite hierarchies we've relied on in the past. Let's get started with defining what makes a hub unique, how to create one, and some rules on when hubs make sense and when they don't.

Site templates and the role of hubs

We've looked at **Team Sites** and **Communication Sites** as modern templates for building new sites in SharePoint Online. Each site is something of an island unto itself. People create content for themselves and store it in places they have access to – let's say one giant treasure chest. Sometimes, several groups of people work together in such a way that they need to start putting boxes, dividers, and containers within that chest and only hand out keys to the people who really need access to each part. We're building what was traditionally referred to as a **site collection**.

This older title makes sense in a world where a site collection served as a boundary for gathering multiple subsites together into a single container. For example, a **Human Resources** (**HR**) site collection may have had a benefits subsite, a policies subsite, a subsite for managers to coordinate, and so forth. The subsites could either inherit the permissions from the parent HR site, or they could override it and choose to change who has access.

The site collection also defined a boundary for how to roll up information. Site collections were contained within their own database on-premises, so that makes sense from a data availability perspective. Web parts such as the **Content Query Web Part** (**CQWP**) provided a way to aggregate content from all sites within that hierarchy into a single result set.

The information architecture needs are the same now as they've always been in SharePoint. We need to define an outer boundary across multiple places where people are working on content and have those areas work together in a coherent way. Subsites were a good answer at a different time but a site collection was a silo with multiple subsite compartments inside. It was difficult to relocate subsites and have them easily connect to a new silo. Hub sites are a much more flexible modern option since spoke sites retain their individuality and can easily change association by leaving and joining another hub.

Defining hubs

Let's define what a hub site is. Usually, only modern sites are configured as hub sites, since that was the original implementation, but any site can become a hub by being configured in the SharePoint Online admin center (shown in *Figure 7.1*). A hub is a site family, and the hub site is the topmost parent in that family. The home page should always be a modern site page to see the benefits of the site becoming a hub.

By registering a hub, we are stating that a site serves as the origination point, home base, or parent to all other sites (once again created with any template) that may be associated with that hub. When a site joins a hub, there are some immediate benefits to joining that family:

- **Shared global navigation** – Everyone in the family can stay more easily connected to each other by sharing a common navigation at the top of each hub-connected site.

- **Shared color theme** – We all get matching T-shirts. The branding colors defined by the hub site are automatically pushed to any site that connects to the hub and can't be changed. This allows us to create a singular look and feel within the site family.

- **Shared search scope** – When we search at the hub site level, all connected sites are included when we perform a new search. When we're on one of the connected sites, a new layer in the breadcrumb is added to go back up to the hub before backing all the way up to the organization.

- **Rollup of information** – News and Events web parts included on hub site pages can automatically include every site in the hub. Content can be created in one spot and readily surfaced in another.

- **Shared search across hubs** – We can connect multiple hubs together to share searching even more broadly. One hub can be a parent of another hub. Searching could then be expanded to look at both families at the same time. This allows us to support complex, multi-level hubs that still need to work together.

Creating a hub is very straightforward. First, we need to create a site and identify that site as a hub in the admin center. We see that depicted in *Figure 7.1*. We need a name for the hub, which can be different from the site. That name is what will appear as the home link in the hub navigation. **Parent hub association** is set to **None** by default. We'll discuss parent hubs more in this chapter.

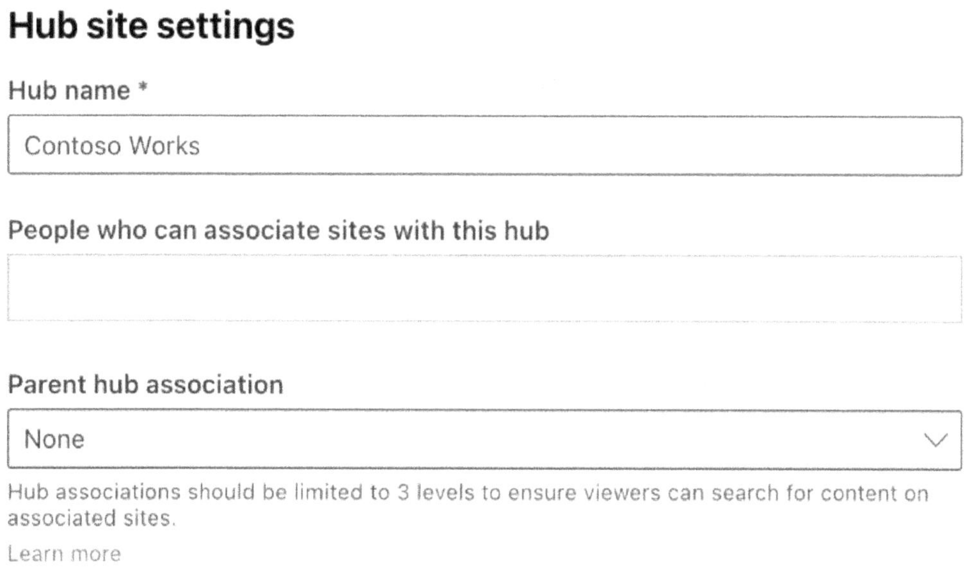

Hub site settings

Hub name *

Contoso Works

People who can associate sites with this hub

Parent hub association

None

Hub associations should be limited to 3 levels to ensure viewers can search for content on associated sites.

Learn more

Figure 7.1 – Hub site settings for a site in the admin center

We can also decide which users who are site owners should have permission to join their sites to the hub. We can leave this open so that a hub can grow organically, or more tightly govern it by restricting who can do the work in their sites. Let's take a look next at why hubs are a preferable architecture to the subsites that have so long existed in SharePoint.

Subpar subsites – the benefits of flat architecture

Site collections define many boundaries in SharePoint on-premises, not least of which was a place to build subsites. Subsites are still an available option in modern sites, but hubs provide a more flexible alternative. What made subsites such a popular option in classic SharePoint? I think more than anything, they were the only self-service option when site creation was otherwise disabled. Even if admins had disabled site creation, users could often still create a subsite within a site collection they were a member of.

In addition to that, it really was much more difficult in SharePoint on-premises to have two site collections share information with each other. Often, the answer was to create subsites in places where they could work with other sites, even if the hierarchy didn't fully make sense. Once a subsite was in place, what if you needed to move it to another site collection?

Aside from a migration tool, the only real options are **PnP PowerShell** or **Content and Structure** under **Site Administration** if publishing features are turned on. Even then, we can move a subsite to a different spot in the hierarchy in the same site collection. Subsites are strict hierarchies. Once a subsite is in a site collection, it's designed to stay there. That doesn't always work for business needs, however.

As an example, if we have a site collection for IT, there may be subsites for development, networking, project management, budgeting, and perhaps a governance subsite that has its own policies and procedures subsite. It might look something like this:

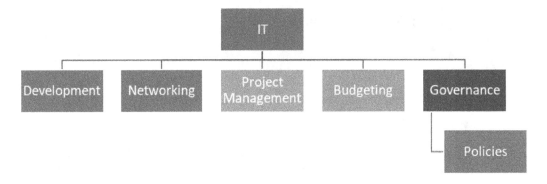

Figure 7.2 – A potential site hierarchy

When the site was created, these connected sub-areas may have fit the organizational structure. But, over time, as the organization grows, we might find that a project management office has grown too and no longer is solely an IT function. Budgeting may become centralized under corporate accounting. Policies and procedures may become a function of knowledge management, and no longer be updated directly by IT. As the organization changes, so must our sites and data. Let's now discuss how where the sites live impacts their flexibility.

Where should sites live?

I like to think of subsites in a site collection like units in an apartment building. The contents inside may come and go, but the unit is there to stay until the entire complex comes down. I may be able to move my stuff to a unit I prefer more, but I can't relocate my unit to a completely different side of the building or a higher floor. The structure is set and is not designed to change.

Hub sites are more like mobile homes. Each one is independent but can be grouped together into a neighborhood to work closely together as a group. If the purpose or the desire of the group changes, the home can join a new community. The site can easily join a new hub. Our previous structured hierarchy now becomes depicted more as a set of relationships between sites, as follows:

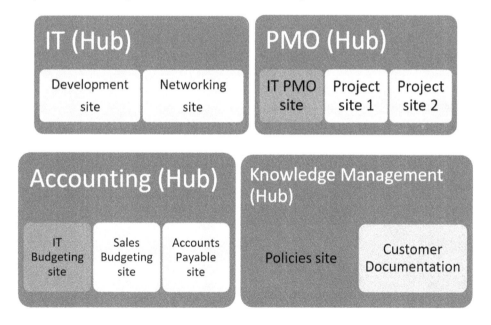

Figure 7.3 – A visual representation of sites existing within hubs

When the structure of the organization or the desired architecture of sites changes in the future, we can easily disconnect a site from one hub and then join it to another. When the hub site moves, its permissions are retained, making it so much more flexible than subsites that had permissions reset to align with its new site collection. We could even have one of our sites that is part of one family leave and form its own. Congratulations proud papa site! Next, let's look closer at what happens when a site joins a new hub.

Hub site benefits – roll up and push down

In an earlier section, we discussed the implications of adding a site to a hub. While the ability to quickly associate, dissociate, and reassociate sites to hubs is compelling, there are other benefits for a site that joins a hub. If sites in SharePoint Online can be likened to homes, then the hub could be considered a **Homeowners Association (HOA)**. Hubs are like a neighborhood.

When I decide to move into the neighborhood, there are certain rules I have to follow. I can't paint my house hot pink. I can't build a 20-foot-tall security fence in my backyard. I have to remain in line with the expectations about how things look and feel. When a site joins a hub, something similar happens. If a color theme is applied to the hub site, it will be automatically pushed down to the sites in that hub. There is no choice but to accept it or leave the neighborhood.

In exchange for following the rules of the neighborhood, I also get to enjoy the benefits of being a part of the neighborhood as well. The streets are cleaned and maintained. The common areas are mowed and cared for. We can visit the pool or the fitness center. Any site in a hub can take advantage of shared global navigation. This appears at the top of the site above the title and site image. We see that in the dark band in the following figure:

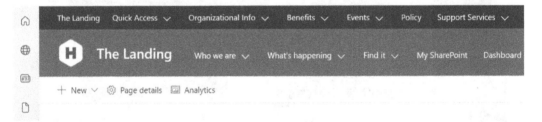

Figure 7.4 – An example of the hub navigation menu

We also get the advantage of sharing a common search vertical, to which we can navigate from the top of the hub or from any of the connected sites. The following screen capture shows a site called Benefits, which is part of a hub called Contoso Works. When we do a keyword search on the site, results are shown from that site, but we can easily navigate up to the hub or onto the organization level and search across all sites:

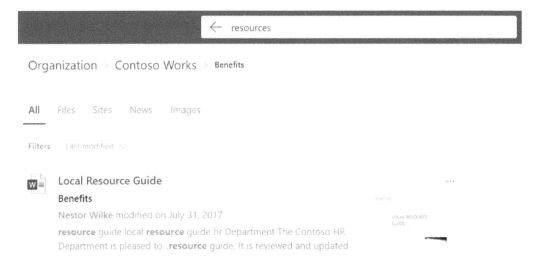

Figure 7.5 – Performing a keyword search on a site connected to a hub with a search nav breadcrumb

The theme, global navigation, and search scope get *pushed down*. We also have an easy and effective way to surface information that may be created on sites within the hub to the top-level site of the hub.

Rolling up – data floats to the top of the hub

Imagine that we have an organization that has grown over the years by acquiring other companies. While there is a broad, corporate identity, each of the acquired companies still keeps its branding, identities, and functions intact. We build a new site for HR at this company and realize that each of the dozen acquired brands also has its own HR functions, and thus its own HR sites. There are no plans to coalesce them into a single HR department, but we do need to have all HR data across the brands work together as a unit.

We could build an HR hub, which would be a site that sits at the top of the hierarchy of the various HR groups within the organization. Each brand's HR site would be added to the hub. While the permissions may be different for each site, such as preventing one HR manager from seeing hiring or salary data for a brand that isn't their own, we still need to be able to have important information from all the sites rolled up to a single spot. All HR sites are part of the same neighborhood.

News and Events web parts allow that rollup to work quickly and easily within a hub. We don't have to give access to all HR managers to edit the main hub site directly. They can create content within their own areas, and the web parts on the hub site can gather them all into a single location. In the following screen capture, we have a News web part that will automatically show any published news posts in the web part on the hub site. These news items are automatically security trimmed but may also be audience targeted if we want only those within certain groups or roles to see them. We can also leverage the filter property to only show news items tagged with metadata indicating they should roll up to the top level:

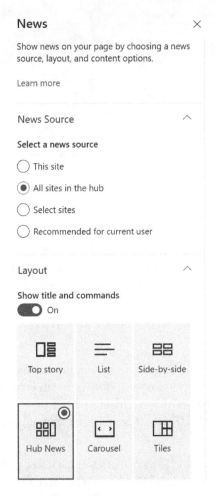

Figure 7.6 – News web part configuration on a hub home page

Hubs are a great way to bring multiple sites together to share information up and down. It allows us to have sites that are not otherwise connected to form a virtual hierarchy. But what if I have two hubs that need to work together?

The extended site family – child hubs

We can't fully have one hub include another hub, but we can connect them for specific purposes and benefits. In the SharePoint Online admin center, we can select a hub site and find **Parent hub association** in its settings. That sounds more like a group of concerned citizens with children coming together to protest, but it is the way we add a child hub. We can also see other hubs that are already associated so we know who else is in the family before we join. Here's the UI for that from the admin center:

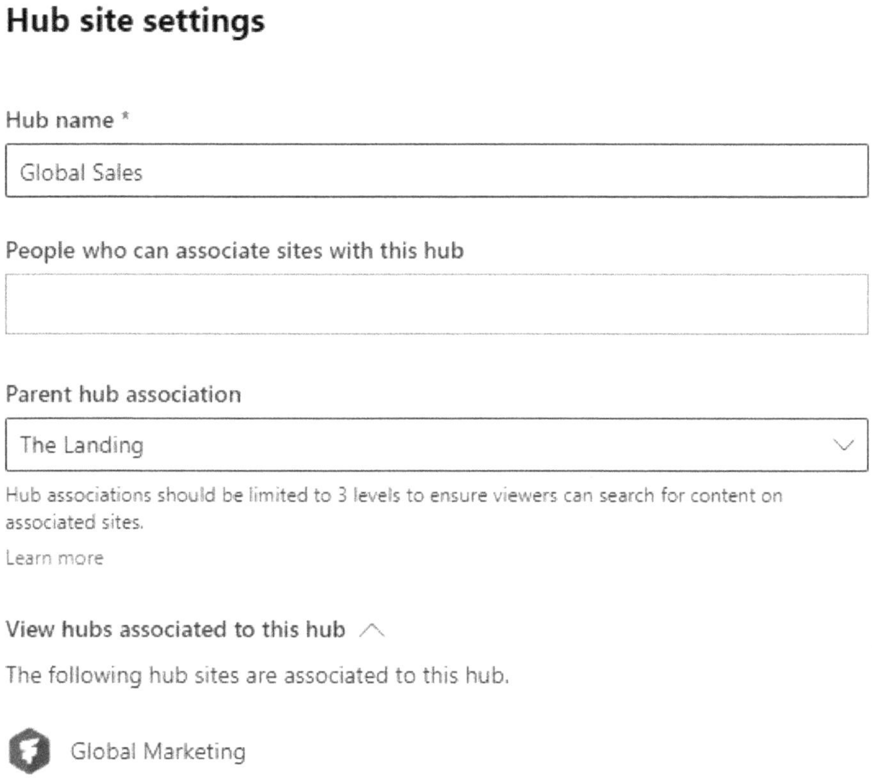

Hub site settings

Hub name *

Global Sales

People who can associate sites with this hub

Parent hub association

The Landing   ∨

Hub associations should be limited to 3 levels to ensure viewers can search for content on associated sites.

Learn more

View hubs associated to this hub ∧

The following hub sites are associated to this hub.

Global Marketing

Figure 7.7 – Adding a child hub in the admin center

So, what actually happens when we associate hubs together? In the example shown in the preceding figure, we have an HR-focused hub that is being connected to the intranet home site. Taking a look at either site will not immediately reveal any differences. Nothing changes in the global navigation for either site. If I search for something on the intranet, I still see the same organization-wide search results because it is a home site. If, however, my parent hub is not the home site, then I see the purpose behind this concept. The child (or nested) hub becomes part of the parent hub's search scope.

What if we also want to depict this hub hierarchy that we're creating in the navigation? Let's assume we have the following structure:

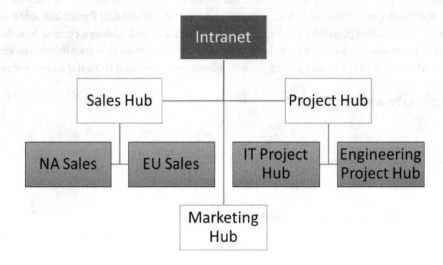

Figure 7.8 – An example site structure

In this example, the intranet is the parent hub. We can choose one of two options when adding a link to its hub navigation, depending on what we want to display, as seen in the following bulleted list. Keep in mind that choosing one of these links just gives you a snapshot in time and adds links that are editable as they would be if they were manually added. That also means that your navigation isn't updated when new nested hubs or sites are added:

- **Associated hubs** – If we're on level 2 or above (the orange boxes in *Figure 7.8*) on the **Sales Hub** and chose this option, links would be added to **NA Sales** and **EU Sales**, the hubs below it.

- **Associated child hubs** – On that same **Sales Hub**, if we chose this option, we would see links to the other child hubs that are connected to the parent hub. In our case, links would be added to the **Project Hub** and the **Marketing Hub**.

It's no coincidence that we're only depicting three levels here. I didn't run out of ink. Nested hubs no longer work (to connect us to search results, which is their purpose) after the third level. There are some other limitations as well. The navigation from the parent hub doesn't cascade down, so there won't be an easy way to navigate back to the top. The color theme doesn't apply between hubs either.

While child hubs may have a place in your tenant for extending search, they may not yet be fully baked enough to become part of your everyday information architecture. Building tall hierarchies between several sites that are complex and difficult to manage is why we left subsites for hub sites in the first place. We still get more flexibility with the option to remove sites from that hierarchy whenever we choose, but a simpler approach may be better.

In this section, we've seen how hubs can allow sites to work together as a voluntary association of sites that should work together. We've seen that the valuable real estate at the top of the site in the hub navigation is also an opportunity to guide our users anywhere, not just within the hub. Let's look at taking that hub navigation to a truly global level.

Hubs and global navigation

One of the benefits that hubs provide is a consistent set of navigation links across the top of all sites associated with the hub. Those links that may be displayed are editable by the hub site owner and are automatically reflected across the entire hub. This may take a few minutes to apply. The only link that automatically gets added is to the hub site itself so that we always have a breadcrumb to get back to the parent home page. We can set an image and title for that link, or we can remove it altogether, as seen in the following screen capture:

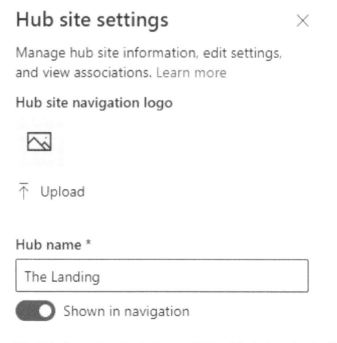

Figure 7.9 – Hub site settings for the logo and title of the hub navigation link

We can have up to 2,000 hub sites in our organization's tenant. We've seen that we can connect multiple hubs to share searching, but the navigation will be different for each hub and within each associated hub as well. While it makes sense to have this flexibility, there may be certain links to sites, systems, tools, departments, or functions that should be available to all users and stay persistent regardless of which site we're on – enter the home site and the SharePoint app bar.

The home site

One lucky communication site in our tenant can be set as the SharePoint home site. Since it serves as the main landing zone of a company intranet, the home site should be the one at the root URL. If there is another site we've been working on and we desire it to be the root, we can perform a site swap. This can be done in the SharePoint admin center using the **Replace** site menu item. Just remember to unregister the site you want to replace as a hub and remove any existing home site assignment:

Active sites

Use this page to sort and filter sites and manage site settings. Learn more

To replace this site, remove it as the home site for your organization and unregister it as a hub site.

+ Create   ✏ Edit   ⚷ Permissions ∨   ⛶ Hub ∨   ⚇ Sharing   ⌂ Change home site   🗑 Delete   ⧉ Replace site

Figure 7.10 – Admin center screen capture of the Replace site menu

We can set the home site in the **Settings** area of the SharePoint admin center or by using this PowerShell cmdlet: `Set-SPOHomeSite -HomeSiteUrl <siteUrl>`, supplying the URL of the desired home site in your tenant. Once the home site has been set, we now have some new options in the **Settings** area of the site itself – global navigation.

Global navigation

Global means everywhere, right? If our tenant is the universe, hub navigation is a guide to the stars in our galaxy. Each hub has its own stars. For something universal, truly global, we need to turn our eyes toward home – the home icon, that is, on the SharePoint app bar. The app bar resides on the left side of all modern SharePoint sites. By default, the home icon takes us to the Microsoft-provided SharePoint home page. We can change that by configuring the global navigation settings shown here:

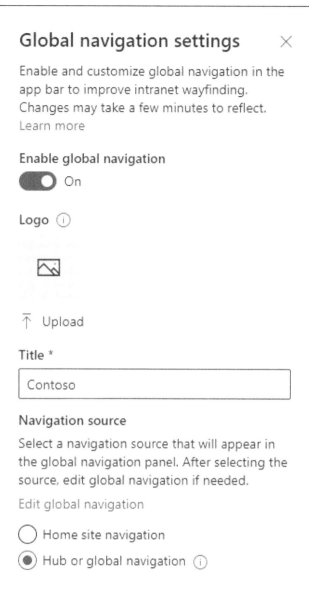

Figure 7.11 – Settings from the home site for global navigation

If we enable global navigation here, we can choose to have the menu that displays when clicking that home icon to be either the site navigation of the home site or the hub navigation if the home site is a hub site. The links will now always display when clicking the home icon, creating global navigation. If we choose the hub navigation to drive the global navigation, it will also display on the home site hub and all associated sites at the top (just like normal). We can also change the title that becomes a header on the **global navigation panel** and the icon to something more in line with our corporate branding.

Global navigation is an important aspect of the app bar, but there are additional components that may help our users quickly find content important to them. We can't edit or control these links, though. They are curated by Microsoft AI for us. Here are the other sections of the app bar:

- **My sites** – Followed or frequently visited sites

- **My news** – News based on your activity and interests as determined by AI

- **My files** – Files in SharePoint and OneDrive you've recently worked on

- **My lists** – Recent or favorite lists driven by the Microsoft Lists service

- **Create** – A way to quickly create sites, documents, or lists

This section has shown us global navigation across all SharePoint sites using a home site, the app bar, and an optional hub. Can it truly be global though if it just exists in one product? It turns out we can expand the reach of our globe into Microsoft Teams as well using Viva Connections.

Viva Connections – bringing the intranet to Teams

The Microsoft Viva platform has been around now for just over a year. Some components have been around longer, living under other names. There are four modules. Products may be licensed separately, or there is a Viva Suite license that provides all of them in a single package. As a suite of tools, Viva is focused on employee experience and engagement, so each tool fulfills part of that promise. Let's explore what the tools are designed to do for us in greater detail. The four components are noted here:

- **Viva Insights** – Personal and organizational insights driven by Microsoft AI to promote well-being and meaningful insights into both good and bad modern work habits. Insights is a Teams app and provides experiences in Outlook and Office.com, including a daily briefing email. Using pre-built queries as well as those provided by Workplace Analytics provides deep manager and leader insights. Personal insights are free but additional features will require an Insights or Viva Suite license.

- **Viva Learning** – A self-service learning option that connects to LinkedIn basic and Microsoft Learn site content for free. SharePoint learning sources may be connected for free as well, though direct support for Learning Pathways is not currently available. Learning is also a Teams app, but content may be surfaced in a Teams tab or private chat. Learning management systems (including Pluralsight, Udemy business, and Coursera) may be connected to Viva Learning with an individual or Viva Suite license. Each person can decide on the topics that interest them or have assignments made.

- **Viva Topics** – Metadata tagging driven by Microsoft AI can be applied to content in SharePoint sites automatically or curated by a knowledge manager. We'll discuss this further in the next chapter. Topics is another paid module with either a license just for Topics or as part of the Viva Suite license.

- **Viva Connections** – A Microsoft Teams app that enables your intranet in Microsoft Teams. Portions of the SharePoint app bar are visible here, as is a Feed web part and an optional dashboard for displaying content with adaptive cards. Viva Connections is included with your Enterprise licensing but needs to be configured.

The last item is obviously the one of greatest concern for this chapter. Viva Connections relies on a communication site becoming a home site and global navigation being configured. From there, our intranet can become an app in Teams that can be deployed to our organization. It is an app that is added in the Teams admin center. When deploying, we can customize the name of the app and the icon that is displayed. The full-color icon must be a PNG file sized 192 x 192. The outline icon should be an inverted PNG file of 32 x 32 pixels. As seen in the following figure, we can also set the accent color of the app by clicking on one of the colors or by providing a hex code or RGB value:

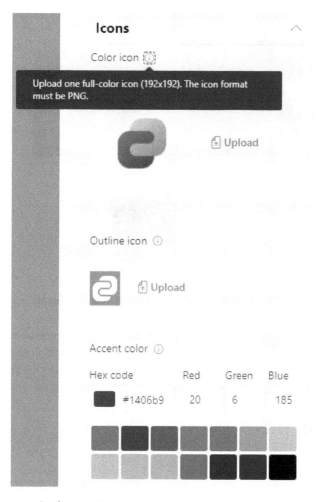

Figure 7.12 – Configuring the Viva Connections app in the Teams admin center

Once the app has been allowed in the Teams admin center, we need to make sure it is made available through an **App permission** policy. Most intranet home sites allow everyone in the organization read/visitor access so it only makes sense to enable the Viva Connections app for everyone in the **Global** (organization-wide default) policy.

Finally, a **Setup** policy can ensure that the app displays in a prominent way for all users, rather than just being made available in the apps list where someone might miss it. Pinning the Viva Connections app to the top will give the intranet a place of prominence and drive both usage and adoption. The following figure shows a sample intranet loaded in Teams using Viva Connections. Take special note of the Teams UI on the left-hand side:

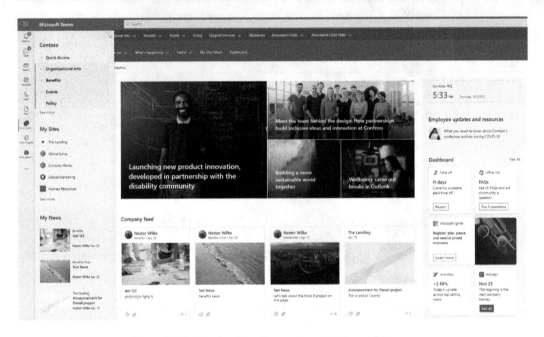

Figure 7.13 – Viewing Viva Connections in Microsoft Teams

Clicking once on **Viva Connections** opens the home site directly in Teams – one-stop shopping! Clicking the icon a second time opens the global navigation (the same one we configured in SharePoint). **My Sites** and **My News** are other components from the SharePoint app bar. On the far right of the page in the preceding screen capture, we also see the dashboard. Let's talk about this and the feed web part in a bit more detail.

Dashboard and feed

The dashboard shows up in a couple of places. First, it is a web part that can be added to our home page. Second, it gets its own tab when viewing Teams on a phone (at the time of this writing, Viva Connections is not supported on tablets but is coming in the second half of 2022) to accompany the **Feed** tab and a **Resources** tab that displays the global navigation.

Wherever it is surfaced, the dashboard is actually created by clicking a **Set up Viva Connections** link on the home site settings. A page is created to hold the dashboard contents. A **View Dashboard** button becomes visible after initial creation. That page can only contain cards designed for the dashboard, not modern SharePoint web parts. Some of the cards are already aware of the data. Other custom cards can be built using the **Adaptive Cards Framework** but the code to populate them would need to be created as an **Adaptive Card Extension** (**ACE**) with the help of *SPFx*.

The following are the cards that are available at the time of this writing. Most cards can be audience targeted so the dashboard becomes focused on different sets of users:

- **Approvals** – Connects to **Approvals** in Microsoft Teams, which can be driven from chat, notifications, Power Automate workflows, or the **Approvals** hub.

- **Assigned Tasks** – Gets data from the **Tasks** app in Teams, which includes **Assigned to me** tasks from **Planner**.

- **Card Designer** – Creates custom cards, which can include text, up to two buttons, and a quick view built using **JSON** and **Adaptive Cards Templating**.

- **Teams app** – A card that links to and opens an available app in Teams.

- **Top News** – This displays news posts that have been boosted by clicking **Boost** on the command bar for a news post. Boosting news makes it more visible, including on this card, until a date that we specify. Up to 10 news posts can be boosted, and we can control their order.

- **Shifts** – Shows user-specific information from the **Shifts Teams** app, including the ability to clock in and out.

- **Web link** – This is essentially a quick links web part, with each card showing text and/or an image and navigating to a different site when clicked.

The dashboard provides an alternative approach to modern web parts in some ways and gives us new ways to connect to data from M365, third-party sources, and our own custom sources and code as well. Since there is overlap between what each delivers, which one do we choose? Obviously, if we haven't implemented Viva Connections yet, the decision is a simple one. For those of us with modern intranets and Viva Connections, perhaps some guidelines we could follow are noted here:

- For links that need an intermediate step or some extra context, we can use the *web link card* in Viva Connections. The intermediate step is the quick view available to custom cards.

- If we want to format the view of a link, we have more UI options with Viva Connections than with the quick links web part.

- For connections to Teams apps, some M365 services such as Planner, or even apps from the SharePoint app store, the dashboard provides capabilities that are not found on modern web parts.

- To implement custom data following a standard JSON format, the custom card can model the data, present it on the card, and connect to it through the code we write.

There is no wrong answer to the question of what to use. Just remember that the dashboard gets a separate, dedicated space on the Teams mobile app and can exist as a separate page or a web part on our SharePoint home sites.

Viva Connections is a free tool that brings our investment into a home site in SharePoint Online into the world of Teams as an app. Teams continues to be a one-stop-shop, and adoption for both Teams and your intranet will benefit.

Summary and planning document

In this chapter, we've explored the concept and implementation details of hub sites in SharePoint Online. We compared their flexible association with the more structured approach of subsites. The shared navigation, search, color theme, and rollup of news and events provide the true benefits of this approach. It is more than just creating a site hierarchy. It's a way of connecting sources of data that may exist in multiple locations, owned by multiple people, and presenting them as a unified site or product that can follow the expectations and user experience of the organization. We also explored how hubs can provide global navigation to a home site and the SharePoint app bar, which forms the foundation for connecting our intranet to Microsoft Teams as an app through Viva Connections.

Here are some items to consider when planning for hubs as part of your SharePoint Online architecture:

- How many subsites exist in your tenant today?

  - Do they need to stay where they are? Does their current ownership and location make sense?

  - Can any subsites be promoted to their own sites?

  - Can any subsites coalesce together?

- How many hubs do I need?

  - Do we just need one for the root of the intranet?

  - Are any business units complex enough to justify having their own site family and thus the creation of a hub?

  - Do any hubs logically roll up into other hubs? Are there parent/child relationships?

  - When does a hub not make sense? Could we use something such as a custom search vertical to roll up information instead?

  - What information should be owned by a group or department, but should also be surfaced on the home page through News, Highlighted Content, or Events web parts?

- What belongs in the global navigation?

  - The app bar should contain the only true global nav? What is important enough to enough people in the organization to take up space there?

  - How many categories or sections should be in the global navigation?

  - Should sections be organized by organizational structure, functional groupings, or some other measure?

  - What links should be global to the hub, even if they are not consistent across other hubs?

  - Should site navigation still be used if there is a hub navigation menu, or will that be too confusing for users?

- Do I need Viva Connections?

  - Is there strong Teams adoption in the organization that would make adoption of your intranet more natural with less context switching if Viva Connections were implemented?

  - Do users interact with Teams on mobile devices? What about your intranet?

  - Are you leveraging custom adaptive cards in Teams or other applications that could be leveraged in Viva Connections as well?

Hubs provide a very flexible way to create and arrange sites. Adding the home site capability allows us to use the global navigation we've defined in a hub for the SharePoint app bar and, ultimately, Viva Connections. The beauty of it is that if the organizational structure or desires change in the future, we can easily rearrange sites in the hub to become part of new site families altogether. While site structure is an important aspect of our information architecture planning, the proper use of metadata with the content in those sites may be even more so. That is where we pick up in our next chapter.

The Mega World of Metadata

We rely on the advantages of metadata regularly but may not even know it. For example, imagine you're making winter travel plans to go skiing. You know your jacket has seen better days, so you decide to hop onto your computer, go to your favorite online retailer, and enter the word `skiing` in the search box. What results will you get back?

Having just entered that keyword, I see that over 10,000 results are waiting for me! To get ready for the trip, we may eventually need clothes, equipment, some lip balm, or even a movie about skiing to get us in the mood, but what we need right now is a jacket. Rather than page through 10,000 results 20 items at a time, we can use metadata to filter our search.

Maybe we choose a department to limit the results to clothing, a favorite brand, a price range, or a condition (new or used). Any time we use information that describes a thing but is not inherent in that thing, we are using metadata. Metadata is sometimes referred to as properties, tagging, attributes, or columns, but the goal is to make storing and finding data easier once the metadata is applied. Sorting, filtering, and grouping all rely on metadata in the world of online shopping or the world of SharePoint data.

In this chapter, we'll explore why metadata is so important, how to implement it in the most beneficial and useful ways, where to create it, and which features and functions it drives in **SharePoint Online (SPO)**. We will go through the following main topics:

- The crucial role of metadata in SPO
- Site columns, content types, and document templates
- Managed Metadata
- SharePoint Syntex and Viva Topics
- Metadata on a mission—retention and sensitivity labels

Let's start by building our awareness and discussing why good metadata is so crucial to the success of our SPO environment.

The crucial role of metadata in SPO

Metadata is simply data about data or additional data that we can use to describe something in greater detail. In SharePoint, lists and libraries use metadata to describe content. For lists and libraries, the columns of data used to describe an item are metadata. **Created**, **Created By**, **Modified**, and **Modified By** are out-of-the-box metadata, but we can add any number of additional columns as well. Effective metadata enables better organization, searching, and classification.

While you can search for items and documents by keyword within the content, we are missing out if we don't take the time to plan effective metadata, which can give you additional ways to organize data and aids in both viewing and searching. Additionally, we can attach extra functionality—such as retention, disposition, classification, and sensitivity—to metadata as well.

Here's a simple example. Let's assume your favorite book is stored as a document in SharePoint. The content on the pages themselves is the data. Other attributes that tell us more about the book are metadata. The title, the price, the **International Standard Book Number** (**ISBN**), and the author are all pieces of metadata. Suppose that we now need to find that book in a warehouse full of books—in other words, that document lives with thousands of other documents in our SharePoint library. What additional metadata could we use to find our book quickly, and maybe even others similar to it?

SharePoint libraries also support the use of folders for organizations. Views use metadata to provide multiple ways to organize. Folders provide a single method of organization. Imagine if the bookstore were actually a bunch of smaller bookstores inside. If our book was published in 1998 and the general subject is science fiction, do we put it in the store built for 90s books or do we put it in the store for science fiction books? Being able to search by metadata allows the book to live in one spot but gives us multiple ways to find it.

Understanding the options

To help us get started down the right path, here is a rundown of the tools and technical options available to us to make metadata shine. We'll dig deeper into most of these areas as we proceed.

Where does metadata surface?

We will find metadata showing up and becoming practically useful in several locations, as noted here:

- **Lists and libraries**: Columns based on SharePoint-provided data types created for a specific list or library. A list or library can also provide default metadata columns and values.

- **Views**: Using the columns of metadata, we can create a dedicated page to show the data sorted, filtered, or grouped by that metadata. These pages can be saved or dynamically set each time we look for something.

- **Document sets**: Like folders, but with metadata that is shared with all documents inside. Can be changed at the top level to cascade down.

- **Content types and site columns**: Reusable site columns may be tied to an overall set of columns in a defined information type available across the site. This set may then be used in many lists and libraries.

- **Inside Office documents**: If a document has metadata, you can edit those properties inside of Word docs, Excel spreadsheets, and so on, or even embed them in the body of the document.

- **Retention and sensitivity labels**: Defined in the Microsoft Purview compliance portal. Applicable as **metadata tags** for documents. Can be used to enforce retention policies or tied to information sensitivity settings.

- **Topic from Viva Topics**: Viva Topics can use **artificial intelligence** (**AI**) as a guiding hand to apply a topic to people and content. This helps relieve the frustration of manual metadata tagging, but the topics can still be actively curated by a knowledge manager. Topics show up as embedded links inside content and provide a rollup of content into topic centers.

- **Managed search properties**: Helps both the *Search* and the *Highlighted Content* web part be more effective. Classic and PnP modern refinement panels can use these as well.

With this, we have a sense of where metadata is visible.

How do we add metadata?

Now, let's see how we go about adding values that make it useful. We can do this in the following ways:

- **Manually**: Tag each document with the correct metadata as it's added. SharePoint provides attention views to keep track of what is missed. This means we can go back and add metadata after content is added as well.

- **Power Automate solution**: Automatically apply metadata values based on custom conditions meaningful to the business

- **Configure library/folder defaults**: Set default values for each folder or library to automatically tag content with as many metadata values as you desire.

- **Configure default labels**: Set a default retention or classification label so that any file dropped into a library automatically inherits the proper characteristics.

- **ShareGate spreadsheets**: We reviewed ShareGate as a migration tool in *Chapter 2, Making the Move – Migration Options and Considerations*. We can use spreadsheets populated with document names and locations from a source and automatically apply metadata via a **comma-separated values** (**CSV**) file when documents are migrated.

- **SharePoint Syntex**: Use Microsoft AI to recognize content in documents, extract important information, and apply metadata tags.

- **Viva Topics**: Use AI and curated topic centers to add a topic to documents throughout your sites.

Metadata is useful but should also be consistent across lists and sites.

How do we reuse metadata across the enterprise?

Let's look at options for reuse instead of reinvention. You can choose from the following:

- **Site designs/PnP provisioning**: An easy method to define a starting point for consistent metadata but hard to push incremental changes, as the templates apply at the time the site is built

- **Content-type gallery**: A content-type hub where you can create content types that get published or republished to multiple sites

- **Using a hub to push content types**: Create content types and columns on a hub site and automatically push to associated sites (requires a SharePoint Syntex license)

- **Term store**: A place for reusable values to attach to columns; any choice column that you use more than once or in more than one site could be added as a term set

- **Search managed properties**: Mapping crawled properties from content to managed properties allows us to use metadata to filter and refine search results through the *Highlighted Content* web part.

I think it's clear to see that SharePoint is built on metadata. It's going to be present whether we take full advantage of it or not. Unfortunately, most organizations never fully include metadata in their content management strategy or SPO implementation. From my experience, that is usually due to either not understanding the benefits fully or not being willing to take the time to add metadata from the beginning or train people in how to do so. It may also be due to the fact that metadata support has grown and changed multiple times over the years, so it can be confusing when trying to choose the best approach.

In the next session, let's start looking at where and how to start applying valuable metadata to our content in lists and libraries.

Site columns, content types, and document templates

Lists and libraries in SPO are made for metadata. Columns that we create to build out the structure of a list and additional properties we can add to a library provide the focal point; it's where our metadata resides. For the remainder of the chapter, let's focus on metadata as it relates to libraries and documents.

Getting started

With any document library, we have a place to store files and folders. Our default metadata consists of **Created**, **Created By**, **Modified**, **Modified By**, and name along with additional metadata, such as file size, whether an item is a record, **Like count**, and others visible in the following screenshot:

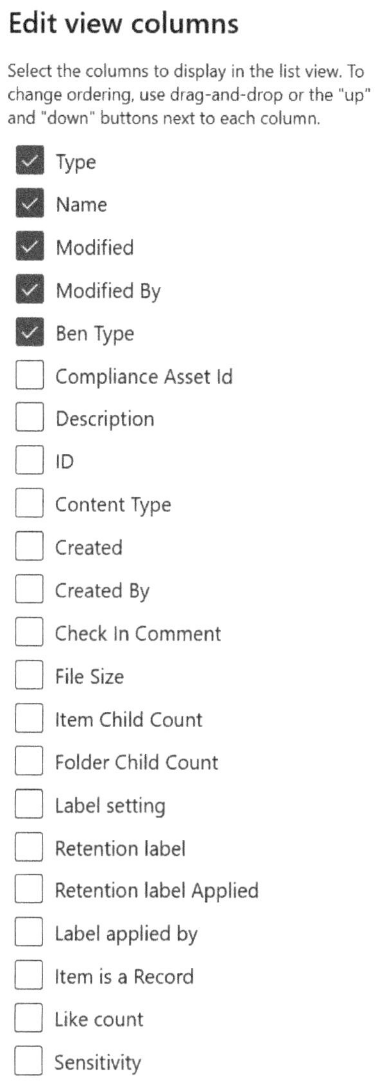

Edit view columns

Select the columns to display in the list view. To change ordering, use drag-and-drop or the "up" and "down" buttons next to each column.

- ☑ Type
- ☑ Name
- ☑ Modified
- ☑ Modified By
- ☑ Ben Type
- ☐ Compliance Asset Id
- ☐ Description
- ☐ ID
- ☐ Content Type
- ☐ Created
- ☐ Created By
- ☐ Check In Comment
- ☐ File Size
- ☐ Item Child Count
- ☐ Folder Child Count
- ☐ Label setting
- ☐ Retention label
- ☐ Retention label Applied
- ☐ Label applied by
- ☐ Item is a Record
- ☐ Like count
- ☐ Sensitivity

Figure 8.1 – Metadata columns as seen when editing a view on a library

While the built-in metadata is certainly valuable, it is through the addition of custom columns that we can leverage our business rules and requirements to provide even greater benefits. This starts with the addition of columns in a library. In a standard view, we can select **Add Column** to create a new column from a predefined list of data types (text, number, date and time, hyperlink, and so forth). We can also add columns by clicking on the gear icon while viewing a list or library and selecting the **List settings** or **Library settings** page. Here, we can see a listing of columns with links to create additional ones for our list or library:

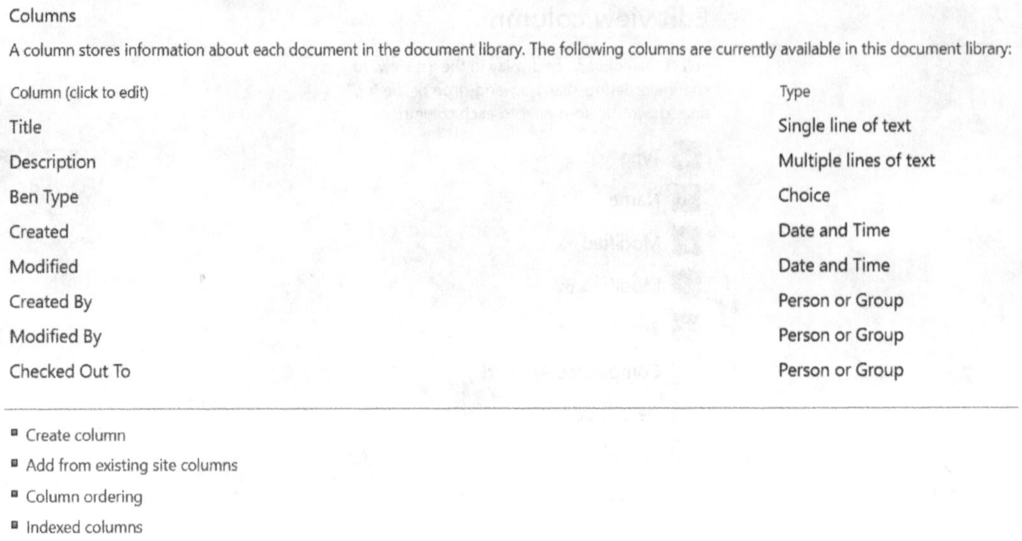

Columns

A column stores information about each document in the document library. The following columns are currently available in this document library:

Column (click to edit)	Type
Title	Single line of text
Description	Multiple lines of text
Ben Type	Choice
Created	Date and Time
Modified	Date and Time
Created By	Person or Group
Modified By	Person or Group
Checked Out To	Person or Group

▪ Create column
▪ Add from existing site columns
▪ Column ordering
▪ Indexed columns

Figure 8.2 – The Columns section of the Settings page for a document library

When a column is added in this manner, we are adding metadata to the single list or library we are currently on. In *Figure 8.2*, there is a custom column named **Ben Type**. This is a choice field created on a library within a **human resources** (**HR**) site and contains a list of available benefit types when adding documents to a library supporting multiple benefit documents for employees (such as life insurance, health insurance, and paid time off). Clicking on the name of the column allows us to edit it after it has been applied, so we could add more choices to the list if needed. We can also mark the column as **Required** so that files in the library will consistently apply that metadata.

Missing metadata

While SharePoint will allow us to fully upload and add new documents to the library without metadata marked as **Required**, we are still reminded that it is needed. If a column is marked as **Required**, we will be asked to supply a value in the details panel after the document is loaded. We will also see an icon to the right of the filename indicating that required metadata is missing and an indication in the view that required information should be supplied, which we see in the following screenshot:

Figure 8.3 – Missing metadata indicators while looking at a library view

Once a document is added that lacks required metadata, a view is also dynamically created for our library, called the **Files that need attention** view. This shows only content that is missing metadata values. These features allow us to clearly see where metadata is missing in each library and hopefully provide an incentive to add it in after the fact. The value of metadata can only be realized when individual pieces of content are tagged with the right set of values to drive sorting, filtering, and searching.

Reusing metadata

Our custom column representing the type of benefit a document is related to may be useful across multiple libraries within the HR site. It could be that our organization comprises multiple sub-entities, each with its own HR sites. So, what if we wanted to use that same list of benefit types in multiple places? We could just keep creating the same column repeatedly, but if changes are necessary later, that's a lot of work for us to update all those locations. Site columns can be grouped together into content types and made available to the search index to become available across sites. Document templates and managed metadata can help us to reuse a document layout or a set of values. The goal is to reuse as much as possible.

Site columns

To reuse the same column in multiple libraries, we need to use **site columns**. These are columns that are created at the site level and are available to use across any number of lists or libraries within a particular site. For example, if we were building an item inventory for a retailer, we may have a list for each product category—so… appliances, clothing, office supplies, food, and whatever else may be appropriate. While each set of items would have its own unique characteristics, there may be some repeatable and reusable metadata we could leverage. All items would have a purchase price, a sale price, and a supplier. For the supplier column, we may have a set list of supplier choices to choose from that are the same across categories and thus across lists. If we create a supplier site column, we can simply add it to each list rather than having multiple, discrete copies on each list.

Site columns must be created in advance by going to the **Site Settings** page. In the **Web Designer Galleries** section, we can click on **Site columns** to create what we need. Several will already be created for us, but we can create our own by defining a column name, data type, and group. I've found it a best practice to always create a custom group for site columns that are custom (rather than using the **Custom Columns** group). This allows us to use meaningful names such as the name of a company, project, or the like.

We may also use a specific prefix to help us identify custom columns when the context of the grouping isn't clear. Where we want to use that column, we can now select it rather than recreate it. We need to do this step from the **List settings** or **Library settings** page by choosing **Add from existing site columns**. When we click that link, we can select our group to easily filter and view only the columns we've created. We can choose whether to add the new column to the default view, as seen in the following screenshot:

Figure 8.4 – Adding a site column to a library

Site columns are not only important for reusability, but also for making search work more effectively.

Crawled and managed properties

The search crawler in SPO is continually looking for newly added content to display in search results and search-driven components. When a new document is added to a library, the content of the file will be crawled and added to the search index. Our metadata can also be automatically added to the search index as crawled properties. Creating site columns can help automate this process.

If we have created a site column and content has been added for that property in a list or library, a crawled property may be added automatically. This will be true if the site column is created in a communication site, or if a **Site Collection Administrator** has added a site column in a modern team site. Search will then automatically add the crawled property as a managed property in the search index, which may be used to drive refiners and filters in components such as PnP Modern Search Web Parts or the Highlighted Content Web Part.

When a new column is added to the search index, it *does not* re-crawl existing content until the documents are edited and re-saved. This is one of the major limitations of the search crawl design. So, metadata planning really needs to take place before content is added to the system.

We can do this work manually by adding crawled properties and mapping them to managed properties in the SharePoint admin center by going to `https://[TenantName]-admin.sharepoint.com/_layouts/15/searchadmin/TA_SearchAdministration.aspx` and managing the search schema. This would require us to use one of the built-in managed properties consistent with the data type of a crawled property, such as `RefinableString00` or `RefinableDate01`. There are a finite number of these managed properties within your search schema, so relying on site columns to automate this work will save time and potential aggravation.

Reusing one column across libraries is valuable, but what if there are several properties that work together and need to travel together as well? Let's talk about grouping columns together with content types.

Content types – packages of metadata

A **content type** is a set of site columns that are defined together under a single label for reuse. Items for a list and documents for a library are built-in content types. We can create our own to encapsulate business data that describes an item or document consistently across lists and libraries.

In our earlier example of a library in an HR site, we had a column that indicated the type of applicable benefit. A content type for a benefits document would allow us to keep related information that describes a document together as a package that we might call **Benefits Document**. It might contain the type/category defined earlier, as well as the applicable year, area of the company where it applies, and perhaps a list of applicable departments.

We can build the content type in the same place we built our site column on the **Site Settings** page. This would make it available to all libraries in a site. If we want to make the content type available across multiple sites, we can build it in the **Content type gallery** found in the SharePoint admin center instead, at this **Uniform Resource Locator (URL)**: `https://[TenantName].sharepoint.com/sites/ContosoWorks/_layouts/15/SiteAdmin.aspx#/contentTypes`. If we create it in the site, it's immediately available. Content types from the gallery will need to be published for sites to use them. Once published, Microsoft has shifted from a *push-everywhere* model to a *pull-as-needed* model. Once a content type has been added to a list or library from the gallery, updates to the source in the **Content type gallery** will be propagated, but only to where the content types are used.

In either place, we build the content type by first creating all the columns (or adding them on the content-type definition page) and then just associate them under the content type we create. We will need to define a category, and we'll need to select a parent content type to start from. This will allow us to define whether it will be available to lists, libraries, or folders within both. The result would look something like this:

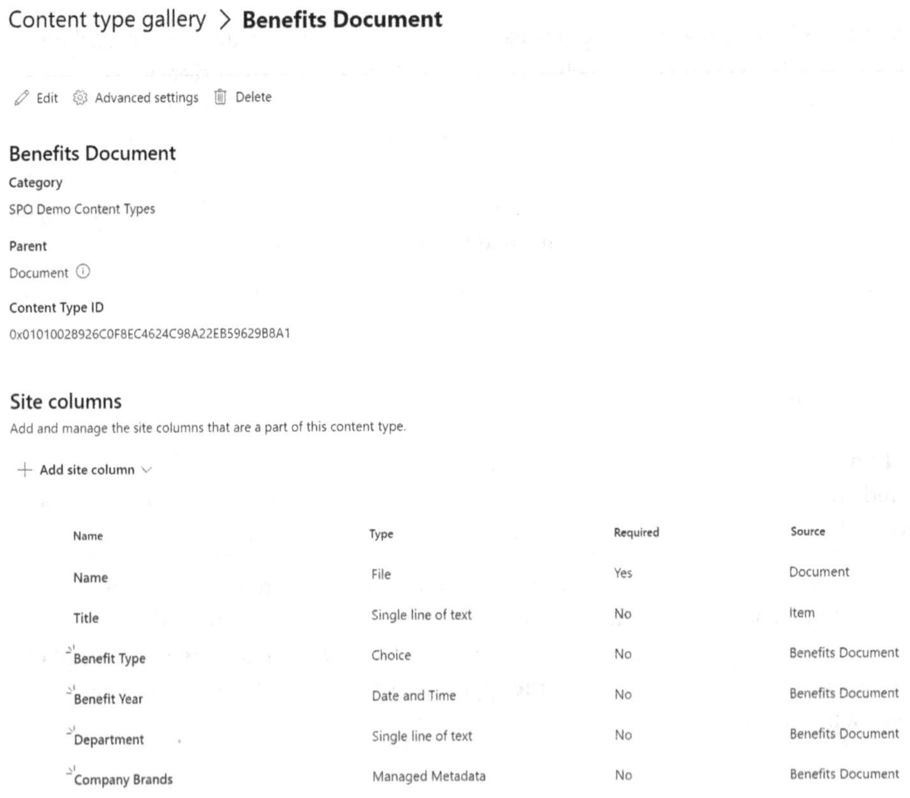

Content type gallery > **Benefits Document**

    🖉 Edit   ⚙ Advanced settings   🗑 Delete

Benefits Document

Category
SPO Demo Content Types

Parent
Document ⓘ

Content Type ID
0x01010028926C0F8EC4624C98A22EB59629B8A1

Site columns
Add and manage the site columns that are a part of this content type.

  + Add site column ⌄

Name	Type	Required	Source
Name	File	Yes	Document
Title	Single line of text	No	Item
Benefit Type	Choice	No	Benefits Document
Benefit Year	Date and Time	No	Benefits Document
Department	Single line of text	No	Benefits Document
Company Brands	Managed Metadata	No	Benefits Document

Figure 8.5 – An example of a content type created in the Content type gallery

By having a benefits content type, we can now have all the columns travel together and be applied to any number of libraries in our site, or across sites if we've used the gallery. We just need to make the library aware that the type exists. We can choose to use it alongside other content types we've defined or with built-in document types such as Word, Excel, or PowerPoint.

We must first enable content types for a list or library from its **Settings** page. We need to go to **Library settings** | **Advanced Settings**, then select **Yes** for **Allow management of content types**. We can continue to use the classic approach of adding our specific content type from that **Settings** page or use the new modern option by adding the **Content Type** column in the list or library view directly. By clicking **Add column** on the view and choosing **Content Type**, we are presented with a screen allowing us to select the published or site-level content types that are available. These are immediately added to the library. If we upload a file, we can change the content type of the document after it's uploaded in the details panel. This has the effect of enabling the site columns on that document, which can then be supplied with the necessary values. We can then add the site columns to the library view as well. We see that in the detail panel shown in the following screenshot:

Figure 8.6 – Screen showing properties on the detail panel

Clicking the **New** button in the upper left of the screen, we see an option to create a new document using our **Benefits Document** content type. While the metadata columns will be applied, the document will still open with a default, blank page. What if we want to start with some boilerplate that should always be present within the benefits document itself?

Document templates

If a benefits document, in our example, should have some starter content or—optionally—a place to surface metadata properties within the content, we can use a template to help us out. This assumes that the document is an Office file (Word, Excel, PowerPoint, OneNote, or Visio drawing). There are a couple of different ways we can leverage templates.

Option 1 – From the New menu

Under the **New** button menu, there are two options we might see. One is **Add template**. Clicking this opens a **Windows Explorer** or **Mac Finder** window where we can select either a template file or an example Office file (so, either a `.dotx` or `.docx` file). That file is then uploaded and surfaced as a new library content type. The name of the file will determine the name of the template. If we don't need to add content placeholders that map to metadata, this option works fine. For context, there are differences between how document templates work in SharePoint versus Office documents in general. In Office documents, the resulting `.docx` file retains a link to the `.dotx` file. In SharePoint, it makes a copy but doesn't use the `.dotx` file as an Office template.

Option 2 – Creating a modern template menu item

Under the same **New** button, we have the option to create a template. While it starts the same way with an upload of a document serving as the template, we also get an online editor that can be used to insert placeholders inside the document, which acts as a place to surface whichever values are added through the metadata properties. The designer opens the file on the left (which cannot be edited here, so we need to make sure the template is complete before starting). We can highlight text and use the placeholder panel on the right to get things connected, as seen in the following screenshot:

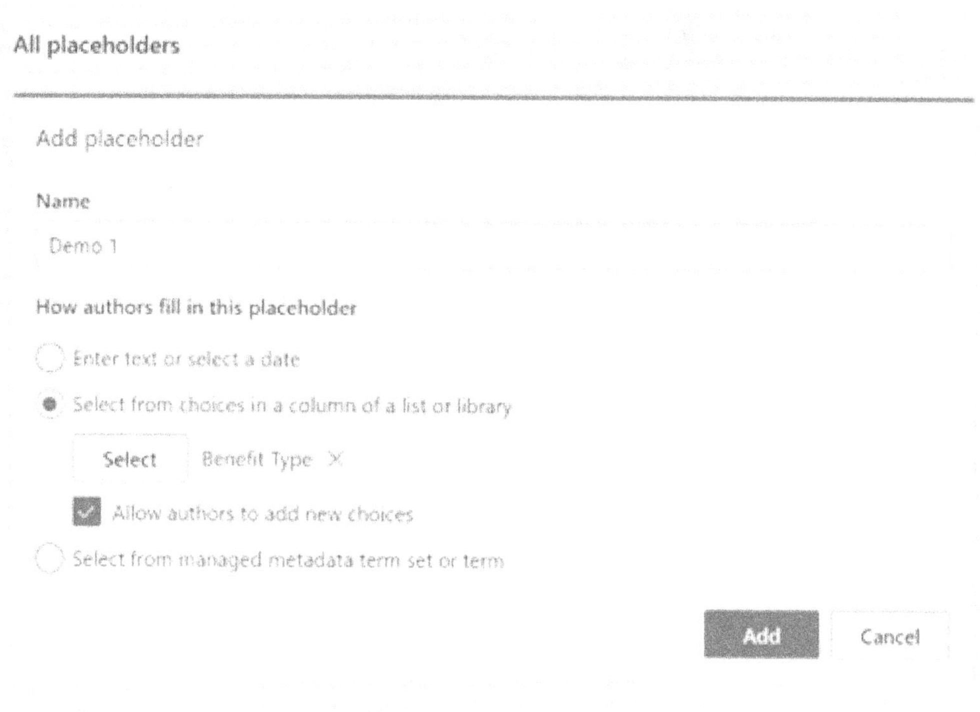

All placeholders

Add placeholder

Name

Demo 1

How authors fill in this placeholder

○ Enter text or select a date

● Select from choices in a column of a list or library

| Select | Benefit Type ✕ |

☑ Allow authors to add new choices

○ Select from managed metadata term set or term

Add   Cancel

Figure 8.7 – Adding placeholders to a modern template

When configuring each placeholder, we can enter text directly, select from column data in a list or library (selecting this will open a dialog to pick a library, then a column), or choose a managed metadata term, which we explore in greater depth shortly.

Option 3 – Adding a template in the Content type gallery

In the **Content type gallery** located in the SPO admin center, we can select a content type that is based on the **Document** parent type and go to **Advanced settings**. From there, we can either supply a URL to a template document already uploaded or we can upload a file to serve as the template from our computer. This does not currently give us the option to add placeholders. After saving changes, the content type will be updated wherever it is currently utilized.

Option 4 – The classic approach in Advanced settings

The modern approaches are preferable, but this is included for sake of completeness. If content types are disabled for a library and we are logged in as owners, we can go to the **Advanced settings** section on the **Library settings** page and either supply a relative path to a file to use as the template or click the **Edit template** link to open the `template.dotx` file for that library locally and make changes.

In this section, we've explored options for creating and reusing individual metadata columns or grouping them together to reuse as a package. We've also seen how content types can have an associated template file. In those content types, we included a **Choice** field. What if that list of choices needs to be included and consistent across multiple content types? That's where **Managed Metadata** comes in.

Managed Metadata

The **Managed Metadata** service has been a mainstay of SharePoint for quite some time. At its core, **Managed Metadata** provides a way to centrally define and reuse sets of terms that can be applied to content as tags or property values. Since the repository of terms can be used consistently across multiple sites, it provides a great way to implement a standard organizational taxonomy.

While site columns create placeholders for values and content types group them together, **Managed Metadata** is where the reusable values themselves reside. Let's review the components that come together to make managed metadata possible. The following diagram shows the objects in the term store and how they relate together:

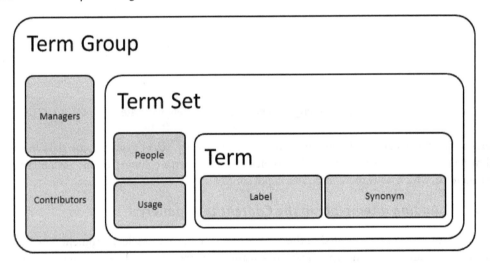

Figure 8.8 – A view of objects and their relationships inside the term store

Let's expand these objects and explore a bit more detail about how they work. In SPO, the experience of managing and using terms has gone through a modernization, though a link to **Return to classic** can be found in the upper-right corner of the **Term store** page. The modern view is seen in the following screenshot:

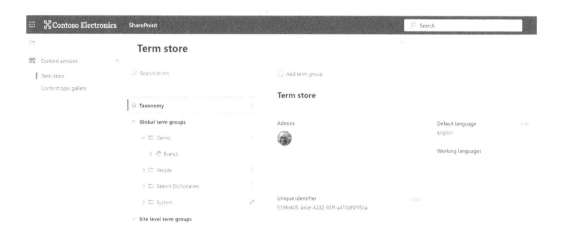

Figure 8.9 – Modern term store

Our starting point is the **Content services** section of the SPO admin center, which includes the following features:

- **Term store**—We get one per tenant. The **Taxonomy** node represents the term store, which gets its own set of admins. To fully utilize the features and functions, we need to be added as term store admins even if we are SharePoint admins. The first global admin is automatically added and is the only one who can add new **Term store** admins.

- **Term groups**—These provide structure for multiple term sets, acting almost like folders. Each group creates a security boundary where group managers and contributors may be added. Group managers can create new term sets and manage permissions for their group but do not have additional admin capabilities. There are two types of term groups, depending on who is responsible for maintaining them, as outlined here:

- **Global term groups** are defined in this global space, and managers or contributors would need to come to admin center pages to do their work.

- **Site-level term groups** are maintained here but are also accessible to site collection admins and site owners from the settings of their respective sites.

- **Term sets**—Each term set is a managed list of terms. Maybe we can think of a term set as a choice field on steroids and more global in scope. To create and manage terms, we must be admins, owners, contributors, or group managers. Terms may be added manually or imported in bulk from a CSV file. There can be a total of 1 million terms defined per tenant. A term set may be copied or moved to a different group. We can also use the following settings:

- **Usage settings**—We can decide whether a term set is open or closed. If open, end users of sites can add their own terms as they go. A term set can be sorted alphabetically or in a custom order.

- **Navigation**—Terms may be used for tagging but they can also be used for navigation. Since there is a hierarchy in a term set, that makes sense but only works at the site level, not for hub or global navigation.

- **Terms and synonyms**—Each term is a value in a listing of values. These values could be hierarchical as well, such as a cascading list of states and then cities. We may have a term set for things such as departments, corporate brands, project codes, or any other organizational collection of data that needs to be used globally and remain consistent. The primary term value is the label. We can also have synonyms. This is helpful when we have *Human Resources*, for example, but also want to call it *HR*. A term may be moved or copied to a different term set.

- **Hashtags and keywords**—We may as well consider this deprecated if we're starting fresh with metadata. The system term set is available out of the box. It contains hashtags and keywords, which are remnants of classic SharePoint. Hashtags could be added to content to make it more searchable and more like social media. Keywords were more like a folksonomy—a perpetual, unstructured list where metadata could be added in a keyword column that surfaced on a per-list or per-library basis. The difference between taxonomy and folksonomy is that taxonomy is a top-down approach to classification and folksonomy is a bottom-up crowdsourced approach. Keywords are no longer added to a list or library by default, but list/library settings may be changed to do so. A system exists along with **People and Search Dictionaries**.

- **Managed metadata columns**—One of the data types we can choose when making a column for a list, library, or site column is **Managed Metadata**. This is really where our term sets have an impact as this is where values are applied, stored, and used for tagging and enhancing searchability. When we click to enter a value into a **Managed Metadata** column, we can click on the tag icon and be presented with a list of possible values. This stays consistent wherever that term set is used.

Managed Metadata provides a way to consistently tag and classify content. I like to think of the example of a grocery store. If I walk into a store I know, I can look at the headings at the top of each aisle and see a term defining what that aisle contains. When entering a new store for the first time, the aisles may be arranged differently, but the labels may still be consistent—produce, dairy, snacks, drinks, and so forth.

Even if the aisles in my store get rearranged, I rely on the consistency of the terms used to find my way, even when locations change. In data that our organization stores in SharePoint lists and libraries, we can use **Managed Metadata** to create a consistent set of values that allow us to find content quickly by searching, sorting, and filtering on the metadata tagging. This example is admittedly one-dimensional. Since terms may be nested, there may be hierarchies where picking one essentially filters the next level down.

One of the biggest challenges is taking the time and effort to build term sets, define content types, and apply metadata at the item or document level. In our next section, we'll look at two metadata tools that may make that process easier with the help of AI.

SharePoint Syntex and Viva Topics

The single biggest blocker to the successful use of metadata has consistently been the amount of time it takes to tag content. This is true for new documents as they are added. That's seen as a burden on busy content creators, and this burden is even greater for content that may have lived in SharePoint for some time. We move from a burden to a nearly insurmountable task.

One shortcut that has helped in the past is setting default metadata values based on a folder structure in a document library. Often, metadata is seen as a replacement for the use of folders since we can sort, filter, and group dynamically based on metadata values. In this case, the two can work together.

On the **Library settings** page, we can click on **Column** default value settings and then choose either the root of the library or any folders that we have created. In the example of a library on an HR site that can contain different types of benefits documents, we can add a folder for life insurance. For that folder, we can set a default choice field value with the right metadata. This is what we see in the following screenshot:

Figure 8.10 – Setting default column values for a column on a specific folder

Now, dropping a file aware of the column into the right folder automatically applies the right value. This can be a very useful option if the folders and metadata are both solidly defined in advance. This may mean more time to set up the architecture, but a more streamlined experience for content owners and creators. There's less manual work for them to do.

SharePoint Syntex

This concept of making the application of metadata easier is where SharePoint Syntex comes into play. The goal of the product is to use AI and **machine learning (ML)** to automate the processing of content and application of metadata (both outside the document and inside the contents, such as with the modern template placeholders previously discussed). Syntex does require a separate license but provides two primary features all geared toward metadata extraction from forms, documents, and images to improve knowledge management and searchability.

- **Document understanding**—This allows us to use unstructured documents with common search phrases to automatically find placeholders in the file, extract them, and automatically populate metadata columns. A no-code model that we create and edit can automatically apply content types and extract metadata columns from Office documents we add to a library enabled with the feature. Training files can be used to mature the model by telling it where to find key information based on search patterns in the document and when it fails to do so. Models can be applied to any number of site libraries.

- **Form processing**—We can leverage the AI Builder tool that is part of Power Automate to create a form processing model. This also requires an additional license for both Power Automate and AI Builder. AI Builder credits are only included with Syntex if you have more than 300 seats. Otherwise, they need to be purchased separately. If form processing is enabled, we can go to **Automate | AI Builder | Create a model** to process a form menu in our document library. From the model, we can extract information from structured files. So, if a contract has the contract number in the upper right, we can use form processing to always locate it. The extracted information may be saved to a new or existing SharePoint list.

These features are configured in the **M365 admin center** under **Settings > Org settings**. A dedicated SharePoint site established as the content center also provides a place for admins to create models and training files (seen in the following screenshot):

Figure 8.11 – The home page of a SharePoint Syntex content center

We can also create additional content centers if it makes sense to do so. In the preceding screenshot, this example is for content used by the HR department. Other **business units (BUs)** could have content centers of their own or use a single instance for all. The goal of SharePoint Syntex is to train a classifier to find metadata for us at the time content is loaded into the system and populate site columns we've created in advance to hold that data.

Viva Topics

This component of the Viva suite exists to bring metadata to content by using AI, ML, Microsoft Graph, search, and the human touch of curated topic pages. With Viva Topics installed, licensed, and running, the discovery process connects and surfaces commonly used terms within your content by associating them with a topic. It also connects people who seem to be commonly creating content or who are strongly associated with a topic.

The limiting factor of lacking time to effectively add metadata is partially addressed here by having Microsoft tag the content for us, but also gives us the ability to guide it and crowdsource the application of it. This use of metadata is in some ways more like enterprise keywords, in that a topic is a single tag or keyword. It can't be joined with others in the same way that we can join site columns within a content type. However, one document could be connected to several topics if they are present within the content. Let's take a look at how topics seem to magically show us the information we need at a glance, how we can configure that magic to our needs, and where to manage it.

Magic topics ride

The magic of Microsoft AI begins to find connections between content and people to create a suggested topic. Viva Topics looks to identify properties and information and displays them on a topic page that exists within the Viva Topics center. The page will include the following:

- Names, alternates, and/or acronyms

- A brief topic description

- People who might know about or be connected to the topic

- Documents, pages, and sites that may be related to the topic

In the following screenshot, we see an example of a suggested topic page for the Blackthorn keyword, which is the name of an executive project related to supply chain in the fictional *Contoso* topic center:

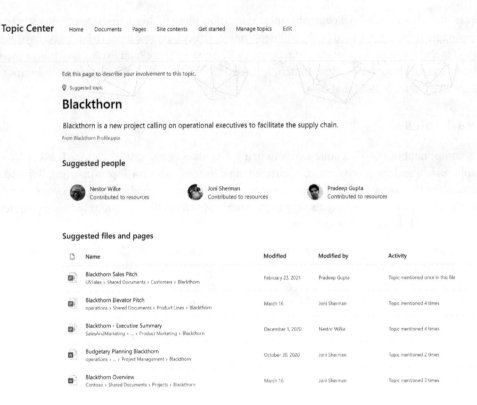

Figure 8.12 – An example of a suggested topic page

Because the `Blackthorn` keyword is mentioned numerous times, a topics page has been automatically generated.

Configuring the magic

We'll look momentarily at how to make the topic page official, but first, we need to understand how we've gotten this far in the first place. In the **M365 admin center | Settings | Org settings**, we find a service link named **Topic Experiences**. It is here that we can configure some parameters around how the **Topics** service should work. These are the options we can configure there:

- **Topic discovery**—Here, we can choose which SharePoint sites should be used to discover topics. The default (and recommendation) is **All sites**, though we can specify a list to include, a list to exclude, none, or a set of selected sites that can be populated by uploading a CSV file. The process of initial discovery can take up to 2 weeks, so it is best to get things started and check back in on the progress. We can also exclude certain people here from being tagged by the AI on suggested topic pages.

- **Topic visibility**—Here, we can define who gets to see topics. We'll discuss where they can be seen momentarily. The default is **Everyone in the organization**. We also have the option to decide if all suggested topics are visible or none. Since topics are designed to be curated by the crowd, **All** would be my choice.

- **Topic permissions**—Only open to those with group admin permissions, we can use this screen to decide who can create new topic pages or edit existing ones. We also have the option of setting who can manage topics in the topic center. This setting provides access to the **Topic management** dashboard. This list should be limited to admins and organizational knowledge managers, who will have the ability to confirm, reject, or view feedback on topics. The resulting role assignments are listed here:

  - **Topic viewer**—Must have a Viva Topics license and visibility to topics but can't make changes to topics.

  - **Topic contributor**—Must have a Viva Topics license and can view and edit topics. These users can also create new topics in the **Topic center**.

  - **Knowledge manager**—Must have a Viva Topics license and can manage topics for the organization on the **Manage topics** page of the **Topic center**. Upon viewing suggested topics, knowledge managers can accept or remove them.

  - **Knowledge admin**—Admins who set up Viva Topics in **Microsoft 365** (**M365**) and have access to manage the **Topic center** and **Topic pages**.

- **Topic Center**—A SharePoint site is created for the **Topic center**. This is defined at the beginning. Only the site title, or **Topic center** name, may be modified.

The Topic center

The **Topic center** is a SharePoint site like any other communication site but built with additional features and capabilities. The home page of the site will show suggested topics and confirmed connections for each logged-in user, as seen in the following screenshot:

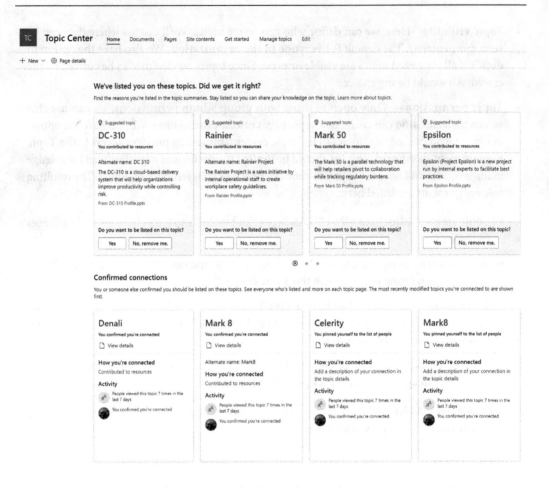

Figure 8.13 – An example Topic center home page

The **Get started** page is one that can be used to plan the rollout and adoption of topics with links to Microsoft adoption and knowledge collateral. The **Manage topics** page is where knowledge managers may view suggested, confirmed, published, and removed topics with a graph to show adoption stats. Topics can be confirmed from the list on the suggested topics page by a knowledge manager.

So, what happens once a topic has been confirmed and published? It starts to show up in a few key locations on our sites.

Interacting with topics

In addition to a visit to the **Topic center** (which may not be the friendliest experience for standard users), topics themselves are surfaced in the following ways:

- **A Topics web part**—If we do want to retain the **user interface** (**UI**) of the **Topic center** but don't want users to navigate to the center's URL, we can include a **Topics web part** on any site. It will display the same information as the **Topic center** home page.

- **Topics highlighted on SharePoint pages and news posts**—If a topic is included on SharePoint pages, it will automatically become a clickable link with a card that pops up on hover. The card will include the title, details, people, documents, sites, and related topics. Clicking the link will take us to the specific topic page in the **Topic center**. The fact that this happens automatically, and no one needs to supply a hashtag or metadata value, is the real power behind Viva Topics. An example is shown here:

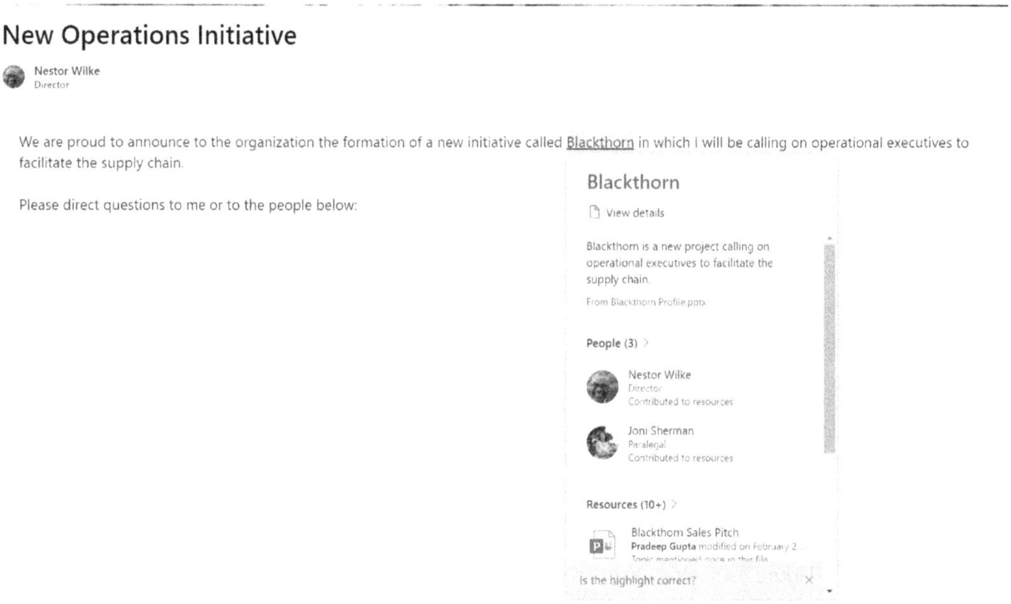

New Operations Initiative

Nestor Wilke
Director

We are proud to announce to the organization the formation of a new initiative called Blackthorn in which I will be calling on operational executives to facilitate the supply chain.

Please direct questions to me or to the people below:

Blackthorn

View details

Blackthorn is a new project calling on operational executives to facilitate the supply chain.
From Blackthorn Profile.pptx

People (3) >

Nestor Wilke
Director
Contributed to resources

Joni Sherman
Paralegal
Contributed to resources

Resources (10+) >

Blackthorn Sales Pitch
Pradeep Gupta modified on February 2.
Topic mentioned once in this file.

Is the highlight correct?   ✕

Figure 8.14 – A topic card visible on a SharePoint news page

- **In Office applications**—In the same way that topic cards are generated and can be clicked on pages, a topic card can be viewed as a result in the search panel on Office documents, opened on the web, or opened using desktop apps.

- **Search results**—If we were to search for the `Blackthorn` keyword, for which we have a published topic, a filter would be visible at the top of the search results page showing the title, description, and associated people, resources, and related topics within the search results page itself. After this, normal search results will display as well. So, a topic gives us a consolidated view across the topic material in one spot within the search, as shown in the following screenshot:

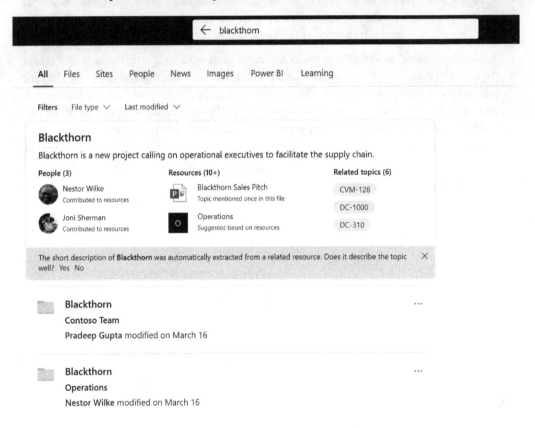

Figure 8.15 – Viewing the Topic filter on a search page in SPO

- **In the Term store**—On a given term set, under the usage settings, we find an option to create topics from terms. Clicking the **Get Started** button provides us with a checkbox list of terms within that term set. We can select each to then generate a new Viva topic for that term. This only works if the term has been applied to content. The term remains and is distinct from the topic if it is successfully created. Changes to the default label or the description of the term will update Viva Topics as well. This allows both metadata approaches to coexist. We can see the details here:

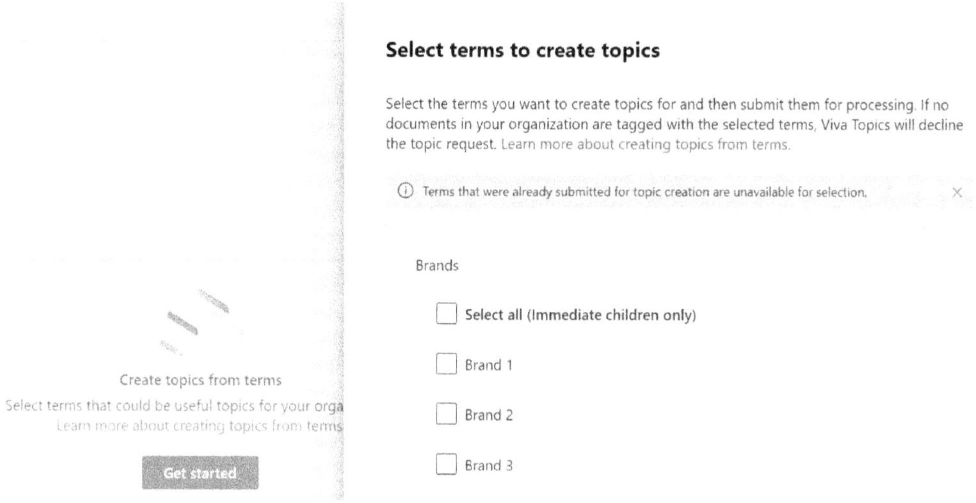

Select terms to create topics

Select the terms you want to create topics for and then submit them for processing. If no documents in your organization are tagged with the selected terms, Viva Topics will decline the topic request. Learn more about creating topics from terms.

ⓘ Terms that were already submitted for topic creation are unavailable for selection.   ✕

Brands

☐ Select all (Immediate children only)

☐ Brand 1

☐ Brand 2

☐ Brand 3

Create topics from terms
Select terms that could be useful topics for your orga
Learn more about creating topics from terms

Get started

Figure 8.16 – Creating a topic from a term in the term store

So, SharePoint Syntex helps us to automate the processing of document information that can be supplied as metadata and surfaced within the content. Viva Topics helps us automate the application of tagging to content based on usage rather than manual effort. Both tools can be effective time savers for content creators and may supplement existing investments in traditional metadata as well.

In the next section, we have one more vantage point to take on metadata. What happens if the metadata value we supply is connected to other processes within M365 such as compliance and retention?

Metadata on a mission – retention and sensitivity labels

So far, we've discussed metadata as a way to drive content management toward consistency for the benefit of filtering, sorting, and searching. We can also use metadata to apply policies to documents that live in SharePoint. The policies can either be retention (how long a document is around and what its disposition rules are) or sensitivity (information protection rules for Office documents in SharePoint and OneDrive).

Retention labels

Retention labels can be defined in the Microsoft Purview (formerly the Compliance center as part of security and compliance) admin center and applied to an item or document in SharePoint or OneDrive, as well as the site itself if needed. Labels are in addition to policies that can be applied to larger sets of documents, such as all files in a particular site. The label is a metadata column that can be applied by an end user, automatically at the library level, or through a document understanding model from SharePoint Syntex.

The label is associated with rules related to how long the document should be kept in the system, actions that should occur when the time comes to delete it, and—optionally—whether content should be locked down as a record. A document may only have one retention label at a time, and the rules take effect when it is applied. The retention label will stay attached to the document even if it moves to a different location in SharePoint but is lost if the content is moved outside the system.

In the following example, we have a retention label configured to keep a document around 3 years from the time of creation and to auto-delete at the end of that time. The assumption is that documents may be uploaded as part of an employee's onboarding that may contain sensitive financial information, and we need to keep the document around for legal compliance reasons:

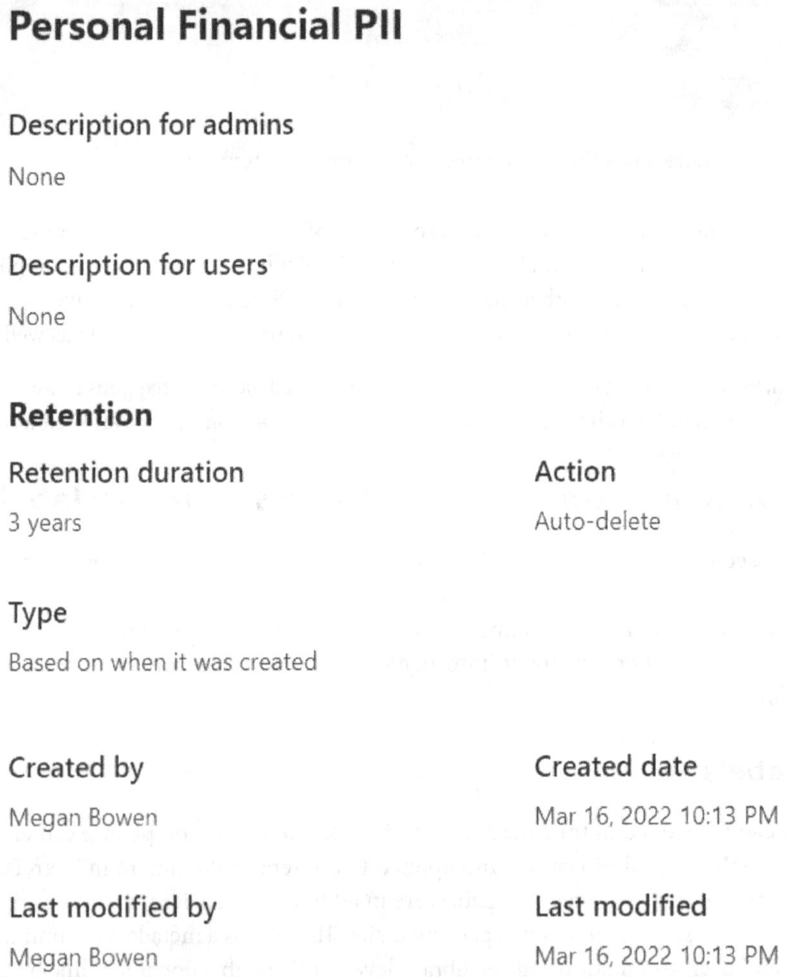

Personal Financial PII

Description for admins

None

Description for users

None

Retention

Retention duration

3 years

Type

Based on when it was created

Created by

Megan Bowen

Last modified by

Megan Bowen

Action

Auto-delete

Created date

Mar 16, 2022 10:13 PM

Last modified

Mar 16, 2022 10:13 PM

Figure 8.17 – Example of a retention label

Once this label is published to either all sites or to a specific site of our choosing (or M365 Groups, Exchange email, or OneDrive for Business), it will become available on the libraries within that site. We can then apply the label we want by name to the column called **Retention Label** (referred to in the details panel as **Apply label**). Even if there are several to choose from, we can only select one. Additional metadata columns automatically capture who applied the label and when it was applied.

If we want all documents in a library to have the same label applied, we can set the default label in the **Permissions and management** section of **Library settings**. This runs when we add a new document automatically, so users don't have to remember to do this. We can also check the box to retroactively apply the label to all content that may already have been created, as shown in the following screenshot:

Apply a label to items in this library

The labels here are provided by your organization to help retain and protect important information. When applied, all items in this library will be subject to the label's settings. For example, if you apply a label that retains content for 1 year, all new items in this library will inherit the label and be retained for 1 year. You can also apply the label to items that already exist in the library.

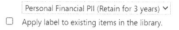

Personal Financial PII (Retain for 3 years) ⌄

☐ Apply label to existing items in the library.

Figure 8.18 – Applying a retention label at the library level

Now, if we attempt to delete a file that is the target of retention, it will appear that we were successful. The document shows in the UI as being deleted. However, in a hidden list viewable and accessible only to site collection administrators and **electronic discovery** (**eDiscovery**) admins, called the **Preservation Hold Library**, our document is retained for the 3-year period with the label specified.

Retention labels are a way to apply different retention rules to varying documents all living in the same place. Retention is concerned only with how long content is kept. For additional access controls, we need to look at the other label option.

Sensitivity labels

These labels are around to help us as we share information inside and outside the organization. They can be applied to sites, groups, and teams or to individual documents, as we've seen with retention labels. Sensitivity labels may be set like a metadata column to a named label, or we may set the label within an Office document as well. These labels are supported on Windows, macOS, iOS, and Android and will remain with the content while inside SharePoint or if downloaded and stored externally.

Sensitivity labels can apply one or more of the following features:

- Encryption and content markings (watermarks, headers, footers)

- Create restrictions on which actions someone could take, such as not allowing offline access or expiring downloaded content after a period of time

- Restrict who can access the file over and above permissions on the library

- Set the default sharing link such as preventing external sharing for documents marked as internal

- Provide an alert when a document containing sensitive information is shared

These additional protections are enabled when the label is applied. We can also use the automatic application of a label so that a user doesn't have to remember to do this.

Both types of labels are added as metadata to a file stored in SPO but have actions tied directly to them. The purpose of the label is not to classify or tag for searching, but to apply other rules and processes.

Summary and planning document

In this chapter, we've explored multiple ways to provide metadata values to documents that reside in SharePoint. List and libraries can have these columns defined just for themselves or we can leverage site columns and content types to work consistently across libraries and/or sites. Adding a document template and placeholders allows us to connect a starter document to a content type so that we can template both the contents and where the metadata connects within the file. The **Managed Metadata** service exists to provide those columns a list of consistent values to keep our metadata usable and up to date across sites in the tenant.

We explored SharePoint Syntex and Viva Topics as tools that allow us to remove the burden of manual application of metadata values and replace this with automatic extraction of that content within the document or generate tags from commonly used keywords. Finally, we found that metadata can also be used to apply retention and sensitivity rules to content that we define in the Microsoft Purview compliance area.

When planning for metadata, the more information we can determine in advance, the more successful our implementation will be. The tools reviewed in this chapter also help make refactoring content to add meaningful metadata for sorting, filtering, grouping, and searching easier and quicker. Next, we'll look at some questions to consider when planning for metadata in SPO.

Sources of metadata

Try to answer the following questions:

- Do filenames currently contain metadata such as a date, project name, or department?

- Is anyone keeping spreadsheets of additional information about items or documents that could be leveraged as metadata?

- Do folder structures found on network file shares reveal a hierarchy of metadata values?

- Can we add metadata and keep folders or use metadata to replace source folders instead?

- What are the most searched keywords?

Applying metadata

Now, see if you can answer the following questions:

- Do we have any repetitive choice fields that could become a term set?
- Who are the knowledge managers in the organization? Are they the same as content creators?
- Do we have any enterprise keywords that are used across departments, locations, and sites?
- Do we have resources with time to create content types and apply metadata values?
- Which tools exist that automate the application of metadata?
- Do we need to apply metadata for new files as they are added or for legacy content as well?
- What are the most common use cases for searching content in our organization?
- What are the content sources that naturally span across multiple BUs?

Retention and sensitivity

Now, have a go at answering these questions:

- Do certain documents need to be retained for legal or compliance reasons?
- Does the organization have a comprehensive retention plan?
- Do content owners need to be notified when the retention period reaches an end?
- Do we need to restrict access to content when shared internally or externally?
- Should offline content be allowed? If so, how quickly should it expire?
- Do we apply labels for individual documents, or can they be grouped together?

In this chapter, we've looked at tagging content across sites. In the next chapter, we'll turn our attention to securing those sites and their content, as well as controlling how and when information is shared outside the organization.

Keeping Things Secure – Permissions, Sharing, and DLP

Keeping your door locked can be important. I recall an incident once, having just moved into an apartment, where I forgot to do so, and my next-door neighbor absent-mindedly walked right into my entrance and had almost made his way into my living room before realizing where he was. I was so taken aback just watching him that I didn't say a word, but it gave me and my family quite a fright. From that point on, I always had a conscious thought of locking the door whenever I came home.

With our sites, lists, and libraries, it can be more than just an awkward moment if the wrong people get access to the wrong data. Making sure that the right people have access to the right content is critical to keeping trust in the tools and in one another. When we do decide to let others past our front doors, we provide a set of keys that gives them access. Once they're inside, though, they still may not have free access to everything. We may still lock our bedroom door or our medicine cabinet to make sure our most sensitive content stays safe and secure.

In this chapter, we'll explore how SharePoint permissions can provide the proper access to the right set of individuals or groups. We'll also explore best practices around sharing content internally and externally by digging into the following topics:

- **Microsoft 365 (M365)** groups, **Azure Active Directory (Azure AD)** groups, and SharePoint permissions
- How to control access to sites and content
- Best practices and concerns for sharing content
- Using **data loss prevention (DLP)** to manage external sharing

Let's start by laying out our basic building blocks for permissions in SharePoint.

M365 groups, Azure AD groups, and SharePoint permissions

Before we talk about the implications of M365 groups on SharePoint permissions, let's outline some definition around how permissions work in our sites. SharePoint manages access by groups defined within the service itself. These groups are mapped to a granular set of permissions that indicate what can or can't be done by someone who is part of that group. While it is possible for a person to be granted direct permissions to a site, adding them to a group with defined access is preferred and best practice.

These groups are maintained at the site level. This means that when we used subsites, it was really the site collection where all our groups were actually kept, even if a group was only used on one subsite. With hub sites, each site maintains its own unique set of permissions so that there is no dependency on a site needing a hub to keep track of its access lists. The boundaries of permission groups are site-level, and those permissions cascade down to all nested objects such as libraries and their folders and files, or list items and their attachments. We can break that inheritance to introduce unique permissions at the object level.

Let's continue our discussion with a look at how SharePoint handles security groups internally, how hubs add an opportunity to partially share permissions across sites, how M365 groups impact team sites, and how we can make our M365 groups and Azure AD security groups work together.

SharePoint groups

Modern SharePoint sites provide three groups as starting containers for people who are part of the site—owners, members, and visitors. When we click on the gear icon and go to **Site permissions**, we see these documented as **full control**, **limited control**, and **no control**, as in the following screenshot:

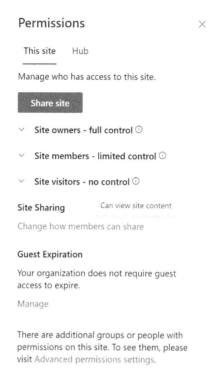

Figure 9.1 – Viewing the permissions of a site

Site admins may add people to these groups. Those in the **Site owners** group have full control over the site, which means they can manipulate these groups as well to add or remove people from the site. In general, we can consider **Site members** as a group of people who have read and write access to all content on the site by default. They can't manage permissions, but they can do just about everything else. Those in the **Site visitors** group have no control, but they do have read access to all the content by default.

Hub site permissions

Visitor access is the one aspect of hub permissions that can propagate down to all sites connected to the hub. If we are on the hub site, there is an extra tab for the hub on the **Permissions** panel. If we enable the **Sync hub permissions to associated sites** toggle, we will be prompted to enter up to 10 individuals or groups who will have visitor access to the connected sites. One of the options here is to use the **Everyone except external users** token to allow all licensed users to have immediate access. This does not change the visitors group on the hub site, so we would need to repeat any permission assignments there.

Owners on the sites associated with the hub would need to accept this setting or choose to manage their own visitors group uniquely. When selecting a hub to associate with, the site owner must choose whether to sync hub permissions to their site, as shown in the following screenshot:

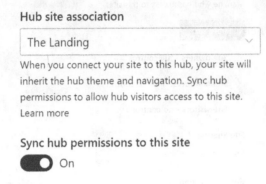

Figure 9.2 – Site information: hub site association for visitor access

The implication of taking this step is that a new group is created called **Hub Visitors** with read access. If we expand the group, we won't see a list of users, but this is how access is granted from the hub. If we set the **Sync hub permissions to this site** toggle in *Figure 9.2* to **Off**, that group will be removed. This allows the site to retain its own set of visitors if needed and to avoid any impact on the people we might assign there.

Team sites versus communication sites

A site owner can add people to the site by going to **Site permissions** from the settings gear. On communication sites, there is a button labeled **Share site**, while on team sites, there is a button labeled **Add members**. The difference in verbiage is significant. As we learned in *Chapter 4, Understanding M365 Groups as the Foundation of Collaboration*, one of the primary differences between the two modern site templates is that team sites are connected to M365 groups. This changes how permissions are mapped.

For communication sites, which do not have an associated M365 group, we can share the site directly. There aren't any other connected resources to be shared as well. When an owner shares a site, they can add an individual account, an M365 group (thus adding all of its owners and members), or an Azure AD security group.

As they are added to one of the three SharePoint groups (owners, members, or visitors), they are effectively being assigned to a permission level, the naming scheme of which is different than the groups. We'll explore permission levels more in a moment, but for now, the owner can add someone with full control, edit, or read permissions, which will add them to the corresponding group noted above.

For Teams sites, the options look a little different. We see two options when adding people to our sites, as seen in the following screenshot:

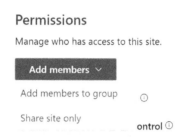

Figure 9.3 – Options for sharing a Teams site

And this is just people. We can't add another M365 group or an Azure AD security group here (but we can bring their worlds together, which we'll discuss in a bit). When we add members to the group, we will provide them with read and write permissions to files and pages in SharePoint, as well as other connected tools such as Teams channels. If someone is added to a group using the Microsoft Teams interface, they will also get access to the SharePoint site and all its contents as a member, with read and write privileges.

The read-only role does not exist in Microsoft Teams. It will never exist for team chats, but if we want to allow for read-only behavior with team files, we have the **Share site only** option. This option allows us to share the site in the same way we can with the communication site template. This is useful if we want to add visitors or add people as collaborators to files in SharePoint without giving them access to other resources such as Teams chat.

Combining types of groups

To bring the worlds together, as previously mentioned, we can rely on group membership to create dynamic Azure AD groups that populate by adding members of other groups using the `memberOf` attribute. This is available in the Azure portal, Microsoft Graph, and PowerShell. This requires a global, Intune, or user administrator and an **Azure AD Premium** license.

We can dynamically populate groups with user attributes such as location or job title. However, these user properties don't always allow us to create the same set of people in an M365 group as we might have in an Azure AD group.

Now, when we set up the dynamic group, we can have its membership populated with users from other Azure AD or M365 groups. For quite a while, we had to potentially duplicate membership in an M365 group to add people to a team or team site, or at least set up some dynamic group filtering that could get us close to the same results. Now, we could simply have an M365 group be forever updated with changes to a security group as its membership changes.

Now that we've explored how to share our sites with individuals or collections of people, let's take a look at how granular we can be with what they are able to do.

How to control access to sites and content

When we add someone to a group in **SharePoint Online** (**SPO**), that group has been configured to allow or block specific types of interactions. These granular permissions are set in named permission levels, and that is our next focus.

Permission levels

A person added to the members group can read and write data in lists and libraries. They can edit pages and create new lists, but they can't add other people to the site. These permissions are set in the **Edit** permission level, which we see as owners when we add people to our sites. We can view the particulars of the **Edit** permission level— and others—if we visit the **Advanced Permissions Settings** page at `https://<<tenant name>>.sharepoint.com/sites/<<siteName>>/_layouts/15/user.aspx`. This page is still in classic mode but offers some deeper insights into SharePoint permissions.

If we click on the **Permission Levels** link in the ribbon, we see there are five of these, even though there are only three groups visible (seen in the following screenshot) in the modern **user interface** (**UI**) for adding users. If we use classic templates such as a **Publishing Site** template, we will see even more:

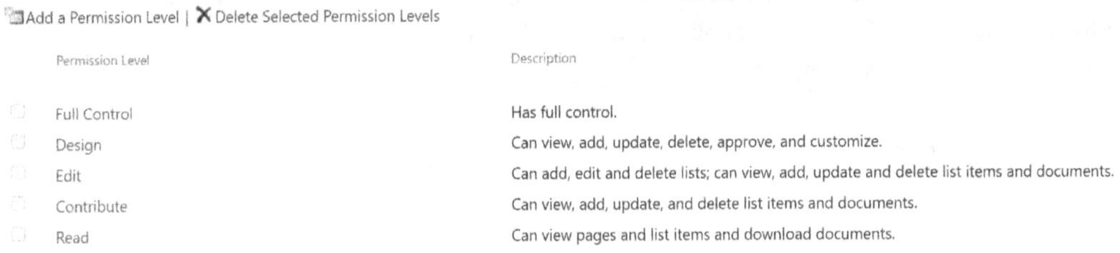

Add a Permission Level | ✖ Delete Selected Permission Levels

Permission Level	Description
Full Control	Has full control.
Design	Can view, add, update, delete, approve, and customize.
Edit	Can add, edit and delete lists; can view, add, update and delete list items and documents.
Contribute	Can view, add, update, and delete list items and documents.
Read	Can view pages and list items and download documents.

Figure 9.4 – Viewing permission levels for an SPO site

A permission level is comprised of three sections:

- **List permissions**: What a user can do with list objects and data
- **Site permissions**: What a user can do to manage the site
- **Personal permissions**: What a user can do to manage personalized views and web parts

A full list with explanations can be found here: `https://docs.microsoft.com/en-us/ sharepoint/sites/user-permissions-and-permission-levels`. Out of the box, **Full Control** is mapped to the owners group, **Edit** is mapped to the members group, and **Read** is mapped to the visitors group. **Design and Contribute** permissions are available as well (remnants from classic solutions), but not mapped to a group. If we use a template other than the two modern templates, we may also see additional permission levels created.

While **Contribute** is not in use for members in modern sites, it's worth calling out how it differs from the **Edit** permission level. The two are almost the same, with one key difference. In **Edit**, this box is checked: **Manage Lists - Create and delete lists, add, or remove columns in a list, and add or remove public views of a list**. So, that means that if someone is a contributor, they can only work in existing lists and libraries, not create their own. In an M365 Groups-backed site, we can't change the assigned permission levels, but we can do so in other modern templates. That may be the better option for many members who are content creators, allowing owners to be the ones who manage the structural containers in a site.

As admins, if we do prefer to map the members group to the **Contribute** permission level, we can go to the **Advanced permissions settings** page and select a group. This should enable the ribbon item to edit user permissions. We see the following screen when we click there, which allows us to change the permission level to either an out-of-the-box selection or one of our custom creations:

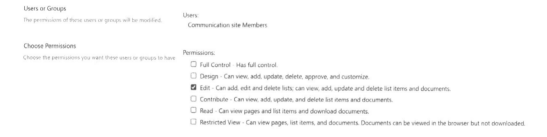

Figure 9.5 – Changing the permission level for a SharePoint group

We can create custom permission levels as well. Our best practice is to create a new permission level and implement our changes there rather than modifying the ones from the system. We can do this easily by scrolling to the bottom of an existing one and clicking **Copy permission level** and saving our own. Some examples of when this may be preferred are if we want to allow people to create and edit documents, but not delete them, or perhaps we want to allow members the ability to approve documents as well as edit them. Custom permission levels allow us to accomplish this.

Requesting access

We may have specific permissions mapped to the SharePoint groups for our sites and people have been mapped to those groups, but those needs access may change over time. To help owners keep up with those user-mapping changes, we can enable access requests so that those without permissions can be prompted with a screen allowing them to ask for it, including an optional comment or justification. We see that setting depicted here:

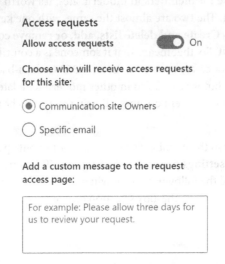

Figure 9.6 – Enabling access requests with notifications sent to all owners

This setting allows people to make requests to see content they don't have permissions to view, including the site as a whole. Site owners can configure who gets notified of access requests and subsequently accept or deny those requests. These requests will include site-sharing attempts by anyone other than a full control user. These automatically go to the owner to process.

If the setting to allow access requests is off, someone trying to access the site will get a standard **Access denied** error page. If on, which is the default, we can specify an email address or have the request route dynamically to the right person(s). For classic sites, communication sites, and modern team sites without a group, that request goes to the users in the owners group. For group-connected sites, the group admins receive the request. If there are multiple owners or admins, the first one to reply has the final word.

Access to team or other M365 group resources is not automatically granted. However, if a member of a Microsoft team requests that someone be added to the team, they do get access to it and the SharePoint site by being added as a member of the group.

Breaking inheritance

An extremely rich but long-forgotten uncle recently passed away and left you his castle and millions of dollars. OK—wrong kind of inheritance! In SharePoint inheritance, files, folders, libraries, and sites all share permissions defined by adding people to groups. So, if someone gets added to the members group, by default they can view and edit data everywhere. There may be instances, however, where we want some content within our site to have a unique set of permissions. We may want to have a subset of people view more sensitive information in a library or only allow a couple of folks to edit and make everyone else read-only.

To break inheritance, we need to go to a list or library, click the gear, and find the permissions link for it. By default, the site permissions are applied to the list or library. When we click **Stop inheriting permissions**, our library now manages its own permissions, but what was there previously has been copied in. So, if we want people in the members group to only have read access, we would need to select that group and edit user permissions, which would only impact the group. We can also delete existing users or add anyone with a license to SPO. If we change our mind later, we can click **Delete unique permissions** and restore things to the default.

Be mindful that breaking inheritance introduces limited access permissions on your site. This happens when an item within a site is shared but not the site itself. This isn't a permission we can add people to directly and it can't be deleted. We can see this in a message in a yellow bar that states: **There are limited access users on this site. Users may have limited access if an item or document under the site has been shared with them**. We should also have the option to click **Show users** to see who is impacted.

In this section, we've seen ways for admins and owners to configure permissions and direct the sharing of access to the site. Breaking inheritance is a way to change the granular level of access on certain containers or pieces of content. Sharing those pieces of content with people who haven't been granted access to the whole site or a library is also crucial for members and a core function of the platform. Let's discuss some best practices for configuring and working with sharing.

Best practices and concerns for sharing content

To share is the point, one might say. People who have access to the site can already get to the content, so sharing is ultimately about letting others outside the site have access as well. From the gear icon, we can get to the **Site Sharing** settings, which allow us to choose from one of the following three options for how data can be shared on a site-by-site basis:

- **Site owners and members can share files, folders, and the site**

- **Same as above except only owners can share the site**

- **Only site owners can share the site and the contents** (this doesn't hide the share menu; it just no longer works)

The latter option effectively turns sharing off on a site (except for owners, of course). If sharing is enabled, the sharing experience for files in SharePoint and OneDrive for Business are both determined by the settings we apply in the SharePoint admin center. We can apply a sharing configuration that impacts all sites at the organization level.

Controlling sharing

Sharing outside the organization is a strength of SharePoint in the cloud. Sending your sensitive information and intellectual property to an anonymous user who can share it with anyone else they want is . . . not a strength. In this section, we'll explore how to control the sharing experience with those outside the organization.

Sharing links

There are two mechanisms for providing access. We've looked at adding people to groups, which is more of a direct access approach. Direct access to pieces of content assumes that a user already has access to the site, though they may not have access, or the right level of access, to that content. Sharing links is more designed for sharing individual pieces of content, which can be done with people who aren't in the site or the connected group. There are multiple sharing links that can be configured as defaults for all sites or configured on a per-site basis.

Under the **Policies** section in the SharePoint admin center, we can define rules for external sharing and set the default behavior of sharing links. Let's review each section and the implications of the settings we find there.

External sharing

In this section, we have two sliders that apply settings to SharePoint and OneDrive. The settings for OneDrive can't be more open or permissive than those we set for SharePoint, as we see here:

Figure 9.7 – Configuring sharing settings in the SharePoint admin center

The most permissive level is **Anyone**. I would heartily recommend this level not be used, as anyone with the shared link **Uniform Resource Locator (URL)** can open a document that's been shared with them with no prior authentication or logging of who's done what. **Anyone** links can be forwarded on to literally anyone else, so they are certainly not secure.

New and existing guests allows us to share content with guest accounts that have been added to the Azure AD by an administrator or by simply using an external email to name the recipient. This is safer than being anonymous because the person must sign in or use a verification code to get to the content.

If we use a verification code and Azure B2B is not enabled, the recipient will need to enter the code each time they need access to the content. This means that the user is never added to the Azure AD. If Azure B2B is enabled, we can configure whether to use one-time codes or have the user sign in.

Existing guests is even more secure still as the guest account must already exist before sharing can occur. To make sure that admins control who those external users are, we need to disable the Azure AD organizational relationships **Members can invite** and **Guests can invite** settings.

The least permissive option is **Only people in your organization**. This disables the ability to share externally across all sites. Chances are, in most cases, this is too restrictive. We may keep this setting as the default but allow broader sharing options for individual sites. For example, if our industry is health care or financie, where there's a lot of private data, we may still need to share information with external vendors for specific reasons. We'll look at making those exceptions in a moment.

External sharing may further be limited to a specific set of allowed domains (seen in the following screenshot). We can limit sharing to only be allowed by people in a specific security group. Perhaps we can even create a group dedicated to this purpose so that sales, marketing, operations, and other users who need to work with vendors or suppliers are allowed while others are blocked:

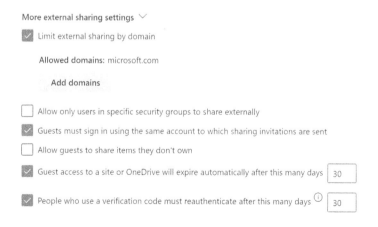

Figure 9.8 – More external sharing settings in the SharePoint admin center

Please note in the preceding screenshot, we have checked the box to have guests sign in with the same account in which they received the invitation and have unchecked the ability to allow guests to share items they didn't create. This, along with the expirations, will help to keep our content more secure.

The other settings we find here allow us to choose the type of default link. With **Anyone** disabled, we can either start with sharing to only people in our organization or by naming specific people. Where sharing with guests is expected, the latter option may be a time saver, but the first is a bit more secure. I would recommend always having the default link permission to be **View** instead of **Edit**, though the latter is the default. Users can change this in either case when they share content.

Site by site

We can also change that configuration to make exceptions at the site level as well. For SharePoint, we can select a site from the **Active Sites** list in the admin center and choose the **Sharing** menu. From here, we can change the selection for external sharing, up or down. For example, we can select **Anyone** even if the default is **Organization only**. A list of domains can be set uniquely here as well. For the default sharing link type and permissions, we can either choose to keep the site settings the same as the organization-level default or change it.

For OneDrive for Business sites, we need to go to the global admin center at `https://admin.microsoft.com` and take these steps:

1. Select an active user.
2. On the **OneDrive** tab for that user, we can navigate to **Sharing.**
3. Select **Manage external sharing.**

If we check the option to allow people outside the organization to have access, we have one of three options to choose, as seen in the following screenshot:

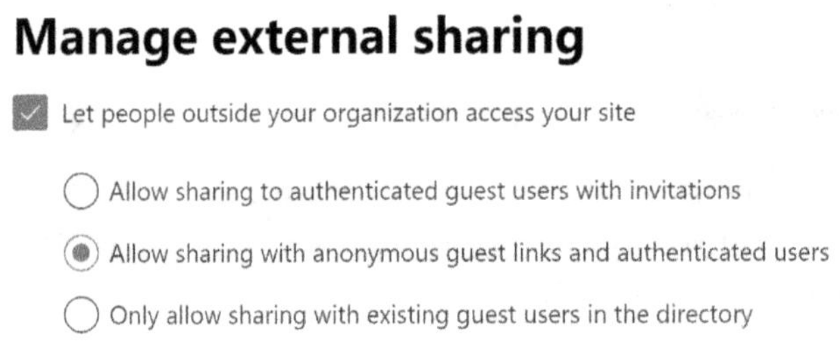

Manage external sharing

☑ Let people outside your organization access your site

    ◯ Allow sharing to authenticated guest users with invitations

    ◉ Allow sharing with anonymous guest links and authenticated users

    ◯ Only allow sharing with existing guest users in the directory

Figure 9.9 – Changing the external sharing options for a OneDrive site

We also have the choice to set these options via PowerShell using the `Set-SPOSite` cmdlet, as follows:

```
Set-SPOSite -Identity https://contoso.sharepoint.com/sites/siteTest
-SharingDomainRestrictionMode AllowList -SharingAllowedDomainList
"contoso.com" -SharingCapability ExternalUserSharingOnly.
```

Taking the time to configure sharing capabilities as early as possible will set us up for a more secure environment. What if we're not opposed to sharing content unless that content has some sensitive data within? The settings we've reviewed so far are the same regardless of what documents contain. If we want to limit sharing based on the detection of sensitive information, we need to also implement DLP policies.

Using DLP to manage external sharing

In the Microsoft Purview admin center, a global or security admin can create and administer policies designed to detect and respond to sensitive information being shared externally. This section is not an exhaustive effort to cover DLP, but an opportunity for us to see how it can enhance and better secure the SPO sharing experience.

Perhaps we can think about DLP policies as performing very specific searches for us in the places we've specified that triggers when it finds something that looks like a match to one of our rules. We as admins and users can be notified and decide how to respond.

Let's break down what DPL policies are and how they are created, and how admins can be alerted if an action triggers a policy.

DLP policies

DLP policies start from templates that fit into one of the following four categories:

- Financial
- Medical and health
- Privacy
- Custom

Each template includes awareness of sensitivity or retention labels, or one or more **sensitive information types (SITs)**. In the first three templates, we have a pre-selected list of SITs. For example, the **United States (US) Financial Data** template includes credit card number, US bank account number, and **American Bankers Association (ABA)** routing number. These SITs are predefined in the admin center, but we can create our own if needed by using **regular expressions (regexes)**. Each defined pattern is accompanied by a confidence level that determines the amount of supporting evidence that is needed per match.

A policy can find and protect data that lives in SharePoint sites or OneDrive for Business accounts, along with Exchange, Teams, devices, on-premises repositories, Microsoft Defender for Cloud Apps, and Power BI workspaces. We define a location when we create the policy. Multiple rules can be applied to a single piece of content, which are evaluated in priority order.

Creating a policy

To perform these steps, we either need to be a global or compliance admin. SharePoint admins can't access the portal otherwise. In the Microsoft Purview compliance portal, we can navigate to the **Solutions** area, locate DLP policies, and create a new one. We can set the following details:

- **Template**: We can search the built-in templates or create one from scratch. The templates will automatically use the sensitive information types they've been configured with.

- **Name**: Replace the name borrowed from the template with a descriptive name for our policy, including trigger, condition, and location information.

- **Locations to protect**: Given our focus on SharePoint and OneDrive, we can either apply our policy to all sites and accounts or pick specific ones. You'll need to enter full site URLs for SharePoint or select from a list of accounts. You can also choose a distribution group to grab everyone in that group. We also have the option to exclude sites or accounts as well.

- **Policy settings**: Choose whether to use the defaults or customize advanced DLP rules that can include conditions, exceptions, options for notifications and overrides, and incident reporting.

- **Protection actions**: We can show notifications and tips in the apps where users are working on content. We can define how many instances of the same sensitive information are needed in an item before the condition is met. We can choose whether admins get emails if a report is triggered or if a DLP rule is matched. The last checkbox enables us to restrict access or encrypt content in the source locations if the policy is triggered.

Here is a screenshot of the possible conditions we can apply if we choose advanced DLP rules:

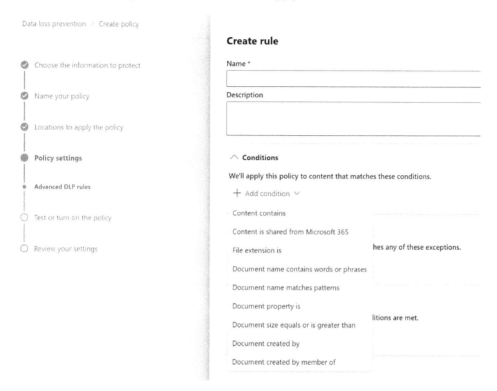

Figure 9.10 – View of advanced rule conditions on a DLP policy

We can join multiple conditions together in the creation of our rule. Note that the **Content is shared from Microsoft 365** condition would allow us to respond if someone tries to share a document from SharePoint or OneDrive for Business (along with email and Teams chat) and that document has already been identified as containing sensitive information.

Sharing alerts

While DLP can automatically prevent someone from sharing a file if it contains sensitive information, there may be times when we just want to receive an admin notification if sharing occurs, not necessarily prevent it. We can do this in the same Microsoft Purview compliance center by creating an alert policy, as seen in the following screenshot:

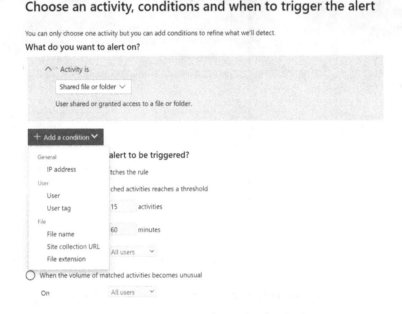

Figure 9.11 – Creating an alert policy for sharing

In the preceding screenshot, we've already chosen **Shared file or folder** as the activity. We could also choose the **Shared file externally** activity if we want to keep close tabs on sharing outside the organization. When the activity triggers, an alert email will be sent to whichever email recipients we've named. We can set a threshold for the number of times the activity occurs within a specified time frame, only get alerted if the volume of activities is unusual, or get an alert every time a file is shared (if we're gluttons for punishment). The condition allows us to only get alerts when a certain **Internet Protocol (IP)** address, user, file, file type, or site is involved.

For end users, we can also leverage email notifications and policy tips to make them aware there is a compliance concern or violation, but still allow them to work. Policy tips can show in documents on SharePoint or OneDrive for Business at the top of the document in a yellow band or as a warning icon that can be expanded for more information.

So, in this section, we've seen that tools in the compliance center allow us to prevent sensitive information from leaving the company through sharing, or at a minimum to let admins, managers, compliance officers, or other concerned stakeholders know that it has occurred.

Summary and planning document

In this chapter, we've explored how to secure and share sites and content in SPO. Many of the concepts are familiar to admins of SharePoint on-premises, but we saw a few distinctions made necessary with the addition of M365 groups. Adding someone to a group automatically adds them to the associated SharePoint site. Sharing only the site with them doesn't bring all the other M365 groups' connected resources along. We also looked at best practices for external sharing and how to both limit it and be notified of it using tools in the Microsoft Purview compliance center.

This information will help us set up the permissions of sites in the most effective way, and to be aware of the implications of M365 groups that we didn't have to worry about on-premises. Since sharing is an extension of the permissions of an item or document outside the organization, we should also have the context we need to configure it well and control/monitor sharing as it happens by using DLP.

Let's define some questions to ask to create the security and permissions section of our planning document.

Permissions

Try to answer the following questions:

- Who should have access to a site? Are they already a part of an M365 group or an Azure AD security group? We should try to avoid adding users directly to a site.

- Should users have access to the SharePoint site only, or to other resources such as a chat in a Microsoft Team, if it's available?

- Are there Azure AD security groups that you want to add to a team or team site that would allow the creation of a dynamic group?

- Should people without access to a site be able to request it, or should they get an error message when attempting to access site content?

- Who are the owners of each site? Should they have full control, or do they really just need to read and edit content?

- Do members need to create their own new lists or just interact with ones that exist?

Sharing

Have a go at answering the following questions:

- Do we want to allow or block external sharing at the tenant level?

- Do we want to enable anonymous sharing, which could introduce more security concerns?

- Which sites should allow a guest to access content, even if it is turned off globally?

- Are there certain users such as field sales reps, marketing, or executives who should be able to share externally from their OneDrive for Business account?

- Should certain people who handle highly sensitive data be blocked from sharing externally?

- Do admins or others need to be notified if documents are shared externally? If so, in all circumstances or just for sensitive content?

- If someone tries to share sensitive data internally or externally, should they be stopped from doing so by using a DLP policy?

Maintaining the security of our data is important and a never-ending process. Communicating the expectations, system settings, and rationale behind security decisions can be complicated if they are perceived to be onerous to the user. Providing awareness of the context and the business needs to keep data safe is critical to getting buy-in from everyone using M365. In the next chapter, we'll explore those concepts and more around successful adoption and effective change management, in security…and beyond.

Part 4: From Current to Change

Part 4 focuses on the user adoption process and experience. This will be less technical and more focused on the human side of change. We will discuss the best practices for adoption with a focus on awareness and knowledge.

The following chapter is included in this part:

- *Chapter 10, The Human Side of SharePoint – Adoption and Managing Change*

10

The Human Side of SharePoint – Adoption and Managing Change

Technology change is human change… every time. This means that we need to account for how people will be impacted by the technologies we use, not just how the technology is configured. In the previous chapters, we focused on the technology and how to effectively plan for how all the pieces and parts work together. Now, we can turn our attention to the human factors of SharePoint Online. You may be wondering why a chapter like this is included in a guide for technical architects.

While it's true that the discipline of change management is on the rise and does constitute an area of focus and expertise on its own, it can't be separated from the other elements of technical implementation of SharePoint Online, **Microsoft 365 (M365)**, or any other technical tool. The process of change management and adoption must be understood and embraced by everyone on a project team – architects, developers, project managers, and change managers alike.

We must all speak the language of change management and advocate for user experience each step of the way. The most thoughtfully designed and prepared intranet just becomes a virtual paperweight if no one understands its value, how to best make use of it, or how it can positively impact their productivity.

In this chapter, we'll explore the deeper meaning of technology adoption, some best practices for driving and guiding successful change, and how to measure and sustain it. To do so, we will cover the following topics:

- Defining adoption and change management
- Preparing for SharePoint Online adoption
- Planning for successful change

- Measuring successful change

- Implementing successful change with ProSci ADKAR

- Sustaining successful change with self-service learning

Let's start by establishing some terms to help us navigate the concepts.

Defining adoption and change management

Change is inevitable, but transformation is not. SharePoint Online and the other collaboration tools in M365 have the potential to have a tremendous impact on the productivity of an organization if they're implemented well and received positively. Adoption and change management concepts are just as important as the technical details in that equation. To understand these concepts, we should probably start by defining these terms.

Adoption

What do we mean by adoption in the context of software and services? Microsoft offers a great starting point for understanding adoption with their definition – "*deep, habitual usage that delivers tangible value to the employees, a line of business, and the organization.*" So, adoption is more than just counting clicks – usage statistics are part of the story, but not the whole story.

We are dog lovers in our house, and all our dogs have been rescue dogs. We've worked with different dog adoption groups, some of which had lengthy and stringent requirements and processes for adopting a pet. While that was frustrating at times, we always kept the perspective that people really care for the dogs and just wanted to make sure that we would be good caretakers. Adoption requires commitment.

This lengthy process probably weeded out those who weren't committed. They made sure we truly wanted to have this dog be a part of our family. They made sure we were knowledgeable about the dog's breed and unique history. Some even visited our home to make sure we had the space, capacity, and ability to care for the dog.

Some of the principles that apply to adopting a dog also apply when adopting new technology tools or new business processes. Change can be hard but starting with a firm commitment to embrace positive and necessary change (no matter how difficult it is or how many times it pees on your rug) is crucial. To summarize, adoption is about usage but also about knowledge and commitment.

Change management

Change and transformation are not the same. Change is a form of disruption, be it good or bad, that demands a response. Transformation is the process of understanding change, embracing it where possible, adjusting to it where necessary, and coming out bright and shiny on the other side. Change management is a set of tools and skills that can help us achieve positive transformation.

When it comes to turning change into transformation with new IT tools, systems, and processes, we generally have two terms that we rally around – change control and change management. While they sound similar, it's important to note that change management focuses on how people are impacted by change, not the process of managing technical updates. They are different approaches to change that serve different purposes.

Let's say, for example, that your car has broken down and you need a new ride.

Change control might help you do the following:

- Diagnose the problem

- Systematically try some fixes

- Keep copious notes about what parts were changed for when another mechanic needs to look later

Change management may help you do the following:

- Navigate the choice of fixing your old car or upgrading to a new one

- Identify what new models and features to consider

- Pinpoint ownership considerations

- Discover how to get the most return on your investment

Change control is about keeping track of what is changing in a system, a server, a software product, or even a process over time. Part of that change control implementation may be limiting maintenance windows or conducting thorough testing in a dev or UAT environment before deploying to production. It may contain a review by a change control board that takes an enterprise-level view of how a change to one system may impact others. There may be logging, comments, notes, and versions, all of which are tracking who changed what and when. While this is incredibly valuable, change control primarily exists to serve IT.

Change management is an arguably broader process that may incorporate change control but is inclusive of awareness initiatives, communication, training, resistance management, and the adoption of new tools and processes. To summarize, change management is concerned with how changes impact the people who experience them.

In short, change management and adoption are always intertwined. Change management is the strategy to ensure successful adoption. It is a set of tools and techniques that enable people to get the greatest value from their technical solutions. Change management helps people adopt technologies that benefit them and are aligned with organizational priorities.

Now, let's take our baseline understanding of change management and discuss how we can apply those concepts to adopting SharePoint Online.

Preparing for successful SharePoint Online adoption

By default, SharePoint and OneDrive for Business are enabled for all users with a license in M365. For Microsoft Teams to work, SharePoint must be enabled. Before your first task can be added to a planner board, a SharePoint site must be created.

Groups have nowhere to put documents without a SharePoint site. In other words, it's hard to avoid using SharePoint if you're using M365. Remember, however, that adoption and usage are not the same.

One of the primary goals of this book is to enable technical architects to configure, govern, and understand SharePoint Online effectively. This is our opportunity to make sure that hard work on the technical side is enhanced by making sure that business needs, user expectations, and productivity impact receive equal attention.

Defining the change and making the case

Why is an organization looking to adopt SharePoint Online? Some common scenarios are as follows:

- To implement a company intranet
- To create a modern file repository solution
- To update an existing SharePoint implementation
- To use it as the target of migration
- To provide the necessary structure to promote Microsoft Teams usage in the organization

Within each scenario, we need to identify the business needs and pain points we seek to address, the most prominent changes that users will experience, and the short-term and long-term plans for measuring success.

If we take the modern file repository as an example, what are the changes that people will experience, and to what benefit? Maybe it is happy to no longer have to support on-premises file servers or manage access through a VPN. For people using the solution, the value may be easier access to files from any device or location or being able to take advantage of co-authoring and version history. The value may lie in having one source of truth rather than having to save multiple copies in multiple locations.

Identifying the change holistically, exploring granular changes that address pain points or common concerns, and establishing the benefits to productivity and experience are great places to begin. Perhaps taking advantage of a simple use case template could help us to define the change and those who will be impacted by it the most:

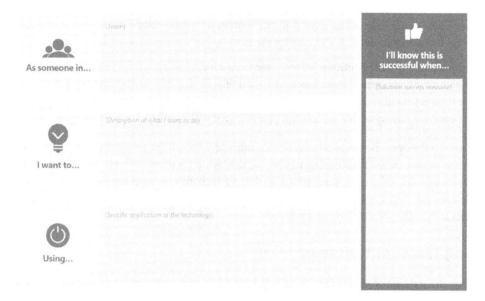

Figure 10.1 – An example of a use case template

This is a standard use case template. It includes the person or persona experiencing the change. We define the future state by what they want to do or accomplish and what tools they may be using to do so. The right-hand side indicates how they know it will be successful. This needs to be specific and measurable so that we can track our progress later on.

Anticipating resistance

With any change, we will naturally have some people who eagerly embrace the change and help lead the way to implementing it, some people who will only accept the change when they must, and some who will resist no matter what. Knowing that resistance is inevitable, we should plan for the best ways to address it.

Resistance usually has one of a few key sources, most of which are rooted in fear and uncertainty:

- **The challenge of learning something new**: It can be an unwelcome task to learn the ins and outs of a new piece of software or service. We may be used to managing data in Excel but don't know what SharePoint lists are or how to get started. This source of resistance can be exacerbated by change fatigue or having too many new things to learn in a short period.

- **Fear of losing productivity**: People get used to doing things the same way repeatedly. It brings a sense of comfort but also helps them do their job more quickly and efficiently. For those of us who already feel overworked, adding a change can seem daunting.

- **Fear of obsolescence**: *How will the change impact my position in the organization? Does the change threaten my expertise, does it make my job less important, or will it cause me to lose my job?* These can be one of the most difficult sources of resistance to overcome and will have to be addressed by leadership and management.

- **Supporting opposing opinions**: Maybe someone is a big fan of some other document management system, but their organization has decided on SharePoint. Maybe the resources that manage the intranet are now comfortable with HTML and CSS and don't want to lose control. Having to become a champion for an opinion you disagree with can be a truly challenging task.

Finding ways to be open, honest, and overly communicative about the impacts of a change can help you avoid resistance rather than having to overcome it, which is often difficult. **What's in it for me (WIIFM)** should drive the communications so that we feel our concerns are being taken seriously and so that we can begin to embrace the changes because we perceive it as being beneficial to us as individuals and to the organization as a whole.

Building the adoption team

Assembling the right set of people from both a technical and user experience perspective is crucial to successful adoption. Identifying stakeholders, sponsors, champions, subject matter experts, and even resisters should be done early in the change and adoption process.

Let's define each of those roles:

- **Stakeholders** should include the core project team that's responsible for implementing a change. We should also include anyone on a **RACI** matrix (a chart of people who are **responsible, accountable, consulted, and informed**), which are in the responsible or accountable roles. Change managers should have a clearly defined and visible role among the stakeholders. They may not be the ones who will execute the changes, but they are crucial to guiding the organization through the change and establishing parameters for successful adoption.

- **Sponsors** should include at least one executive member who is directly responsible for the system, service, or process being changed or who is the leader for an area that will be most impacted. Visible and consistent executive involvement is critical for the change to be taken seriously. They serve as a rallying point to drive home the organizational context and business needs. Other sponsors could include program managers or people managers. Sponsorship should imply ownership and accountability in the change process.

- **Champions** are people who are passionate early adopters – people who can help tell the story of why adoption helps them and the business. They may also be people who are influencers in the organization with relationships that cross departments, locations, or other potential siloes. Champions may volunteer or be recruited by management and leadership. A champion could come from IT, but this role is less about sharing technical details than it is about sharing potential success stories. Influencers, or well-connected people, can also be identified and potentially act as champions for change. Microsoft has a wealth of assets and support opportunities for those identified as champions in your organization: `https://adoption.microsoft.com/become-a-champion/`.

- **Subject matter experts** are the people who know the most about the specifics of the impacted technology or the change process. Given that training and learning are such important components of successful adoption, we should certainly include these stakeholders. For a new modern intranet, for example, these could either be SharePoint admins, content creators who've used similar tools in the past, or trainers who prepare material for formal learning events. SMEs can be considered part of the stakeholder group but may not be involved in the entire life cycle of the change.

- **Resisters** should identify where we expect resistance to come from. This is not about specific individuals, but rather personas that may be impacted negatively, disproportionately, or existentially. Maybe we're migrating files from network shares to SharePoint, and there is resistance we anticipate from sales reps in the field who are comfortable with the current folder structure and are worried about the time it takes to learn something new, taking attention away from their customers. The earlier we account for resistance, the better.

Just like a proper chess board, all the pieces are in place, but now, we have to plan the strategy we are going to execute to succeed. Now, let's look at the five levers of change.

Planning for successful change

Change can and will happen, even without proper planning, but we have all likely experienced a change at work or at home that we weren't prepared for. The chaos around a change in process or technology at work can lead to a bad taste in people's mouths for years, even if the change has ultimately benefited them. Change that is unplanned, poorly planned, or poorly communicated can lead to lower adoption of technical solutions. Low adoption is a result of that lingering bad taste.

Having a consistent model and methodology to manage change is critical to successful outcomes. One of the most well-known and arguably most effective models is from ProSci (founded in 1994 by Jeff Hiatt), which focuses on how individuals handling change leads to effective change at the organizational level. This model is known as **ADKAR**, an acronym for **awareness, desire, knowledge, ability, and reinforcement**.

In the ProSci process, there are three phases – preparing for change, managing change, and reinforcing change.

The middle phase is traditionally where five key planning documents come together so that we can be prepared to go through the change process effectively and peacefully.

Each document is mapped to at least one of the ADKAR phases and is designed to ultimately help each individual with their transition. These documents are sometimes referred to as the five levers of change. These are communications, sponsors, coaching, training, and resistance management.

Communications plan

This plan helps us to decide what information to share, with whom, and at what time. It primarily addresses awareness and reinforcement but can also be used to influence desire. This should take the content and tone of the message into account, as well as the various ways that message might be delivered. Understanding personas is crucial in creating an effective communications plan since the content, tone, and medium may be different for each audience.

This plan should always include a heavy allotment of why the change is needed, what values it will bring to those experiencing it, and what our success measures are. Though a change manager or someone from corporate marketing may write the messages, who it is perceived to come from can be critical. Messages that require authority may be sent from an executive. Messages that require trust may come from someone's immediate manager.

For successful SharePoint adoption, the communication plan may include information on how SharePoint is replacing an old company intranet or how Excel spreadsheets might be moving to lists to support automated approvals. Business units that might be migrating content from file shares to libraries may be given information on how versioning and co-authoring work. Overall, communications around the timeline and the rollout process should be included as well.

Sponsor plan

This plan, also referred to as the sponsor roadmap, should include the detailed actions, expectations, and communications required of senior leaders and/or executives to support the change process. This plan impacts awareness and desire but can also be used to enhance reinforcement.

There may be few sponsors while there's a larger number of stakeholders, but the impact that the sponsors' visible involvement has can greatly increase or decrease the likelihood of success. If a senior leader is enthusiastic about the change, embraces and evangelizes it, and is a consistent presence throughout the change process, employees will be far more likely to believe in the value of the change themselves.

Coaching plan

Managers and direct supervisors experience the change twice, through their own eyes and those of their direct reports. Often, employees have the strongest trust in their peers and their manager, which makes their involvement crucial in successful change.

This plan helps managers do the following:

- Understand how to best communicate the change
- Advocate for the adoption of the new solution, process, or tool
- Coach others on how to embrace the change
- Identify resistors and help address their concerns early
- Be a liaison between their employees, sponsors, and other stakeholders

In a SharePoint-focused adoption effort, managers may be enabled early to understand the benefits of storing documents in libraries where many people can work together. They may have input into the design of their department's home page.

They may be given coaching and tips on how to tell the story of the value of a new intranet, Teams, or process automation that is coming. They can also act as guides and champions to enhance ability and reinforcement. All five elements of ADKAR are directly influenced by managers and supervisors.

Training plan

What will people need to know about the change to be able to find benefit and value in it? Training plans impact knowledge and ability by offering specific, how-to information. It is focused on skills; however, those skills may only make sense if the awareness and desire have been established first. Someone might wonder why they're being given training on how to collaborate on a document in Teams when they don't know what Teams is or why they can't use Slack anymore.

A training plan should include the specific audiences, topics, agendas, and schedule for training. Some training may be presented live (in-person or remote), while some may be documents or links to content on the web. A variety of media usually means greater knowledge as it meets people where they are with their learning styles.

For SharePoint, we often have training for site owners, content owners, knowledge managers, system or service admins, and end users. Cheat sheets or guides with screenshots may also be helpful. Depending on the size of the organization, we may not be able to reach everyone, so live recording sessions are always helpful, especially if there is a Teams or Yammer community where people can ask questions and get answers, even long after a session has been delivered. Too often, training is presented at the beginning of a rollout but is not kept up to date with corrections or new information.

Resistance management plan

Resistance is a feature, not a bug. We should expect it and proactively seek to address it.

Change managers, along with people managers, can work to identify expected areas of resistance in advance and frankly openly address resistance during the change. Meeting people's concerns and reservations, soliciting and incorporating feedback, and just remaining transparent can certainly address desires.

Awareness can also be advanced as we seek to inoculate against change, wariness, and weariness.

The five levers of change allow us to make the heavy lifting that much easier. The communication plan, resistance management plan, and sponsor roadmap help early in the ADKAR process. As we move through the change, the coaching plan can help us (and our manager) to embrace and adopt the change all the way through. The training plan can increase practical knowledge and lead to **ability (the fourth step in the ADKAR model)** if maintained and updated. Now that we understand how to plan, let's learn how to measure the effectiveness of what we've planned.

Measuring successful change

Change is inevitable, but it's not always successful. As our organizations move to embrace the M365 suite of cloud tools, we are surrounded by change. It could be a change in the way we access information, a change in the way we think about structuring and storing files, or just the change fatigue that can arise when software is constantly being updated.

For the successful adoption of SharePoint Online and associated technologies, we need to establish some guidance and governance around the change that people experience to know whether the deployment has been successful and if they have benefited from the deployment as well. While not exhaustive, I'm sure we can start with the following list of success measures to track and recognize success or identify opportunities to improve:

- Quantitative – measuring usage
- Behavioral – measuring changes in how work gets done
- Qualitative – measuring sentiments

Let's take a look at these measures in more detail.

Quantitative – measuring usage

This might be the easiest way to gauge at least one aspect of a successful adoption. Usage metrics tell us how many people are engaging with the technologies and how frequently, and reveal patterns over time. The questions we should ask here are as follows:

- How many people are using SharePoint/Teams/OneDrive/Planner? Out of the total count of employees, what percentage of possible users does this represent?

- What does our telemetry or analytics data tell us about how many times employees open the cloud software or how long they use it each day?

- Are employees still using other tools, possibly unsanctioned by the organization?

- What is the granular activity? How many files, chats, and policies have been created?

The tools we have at our disposal here are as follows:

- **Site usage**: By going to the settings gear in a SharePoint site and clicking **Site usage**, we can see up to 90 days' worth of site usage data for the site we're on. We can view the following information:

  - Number of unique viewers of the site

  - Trends for unique viewers over time

  - The total number of site visits

  - the amount of time spent on modern sites or news pages

  - The most popular content on a site

  - A chart showing the concentration of site traffic by day and hour

  - Whether users are accessing the site by desktop, mobile, app, or tablet

- **Hub site usage reports**: We can see unique viewers and data from hub visits from all connected hub sites for up to 30 days. We can see top pages, news posts, and documents from all hub sites for the last 7 days in the **popular content report**. We can also see which hub sites are the most popular. These reports are not available in China, Germany, or any GCC tenant.

- **Page analytics**: Stats on the number of views, viewers, and time spent on a particular page rather than the site as a whole. We can find these by clicking on the **Analytics** menu link toward the top left of each modern page, which will open the report in the **Details** pane on the right of the page.

- **SharePoint Admin Center reports**: We can view exportable reports here for activity and site usage. Activity reports show the number of files by activity type, the number of unique pages visited, and the number of users by activity type. Site usage reports show us the total number of sites compared with active sites, the same for files, and the amount of storage used. This can be from 7 to 180 days.

- **M365 Admin Center Usage reports**: This shows how users are interacting with a variety of services, including SharePoint, OneDrive for Business, and Teams, along with Forms, Viva Learning and Insights, and other tools. Up to the last 180 days of activity is available in click-through reports, as shown here:

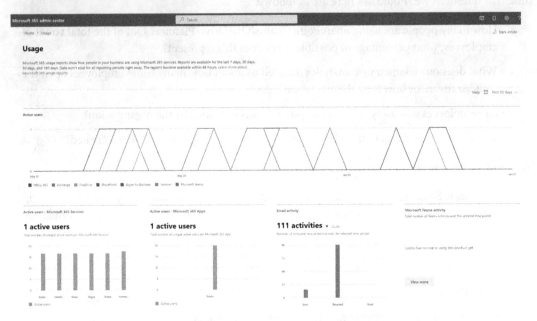

Figure 10.2 – Usage report from the M365 admin center (30 days of activity)

Behavioral – measuring changes in how work gets done

Are changes in behavior, habits, and activities taking place as a result of making SharePoint Online and other cloud tools available and focusing on how they deliver value to those in business? Usage tells us only part of the story. A change in the available technical solutions doesn't always equate to a change in how people work. The questions we should ask here are as follows:

- Using new and better collaboration tools, has the number of emails decreased?

- Are meetings more inclusive of different opinions because we're using tools that allow instant collaboration, such as Whiteboard or Loop components?

- Are requests to human resources for benefits information or requests to IT for service assistance fewer because the information is easier to find on an effective intranet?

The best tools we have to assist in this measurement are Productivity Score and Viva Insights. Productivity Score is a tool in the M365 Admin Center that must be enabled, but within 24 hours, it will show metrics, insights, and recommendations across people and technology experiences, along with comparisons to similar organizations and recommendations for improvement. Productivity Score is a type of quantitative measurement to be sure, but reviewing the score regularly to see the impact that our change management efforts are having on adoption can help us measure changes in behavior over time.

A regular review of people's experiences offers a tremendously useful glimpse into the behavioral changes that are taking place relative to the adoption of cloud tools in the areas of content collaboration, mobility, communication, meetings, and teamwork. We can compare compiled stats from the real activity and compare that to public research done by Microsoft into best practices in each area.

The score itself is a set of eight categories, equally weighted, with the highest possible total of 800 points. Each category includes 28-day and 180-day views of the key activities for that category. The score includes data from Exchange, SharePoint, OneDrive, Teams, Word, Excel, PowerPoint, OneNote, Outlook, Yammer, and Skype and is updated daily. Our organization's score history is maintained so that we can track progress.

The following screenshot shows an example of the top-level view of **Productivity Score**, which shows a score for each person's experience, such as communication modes, effective meetings, online file collaboration, shared workspace collaboration, and the use of mobile devices to work with their team:

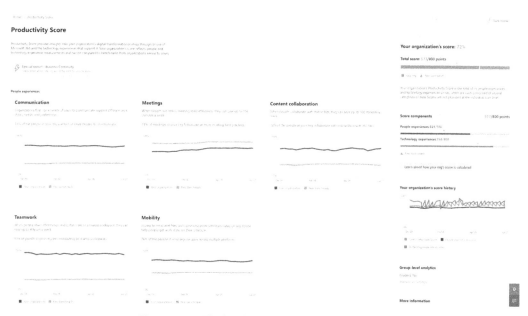

Figure 10.3 – The landing page of Productivity Score

The following screenshot shows a view of only one of the experiences:

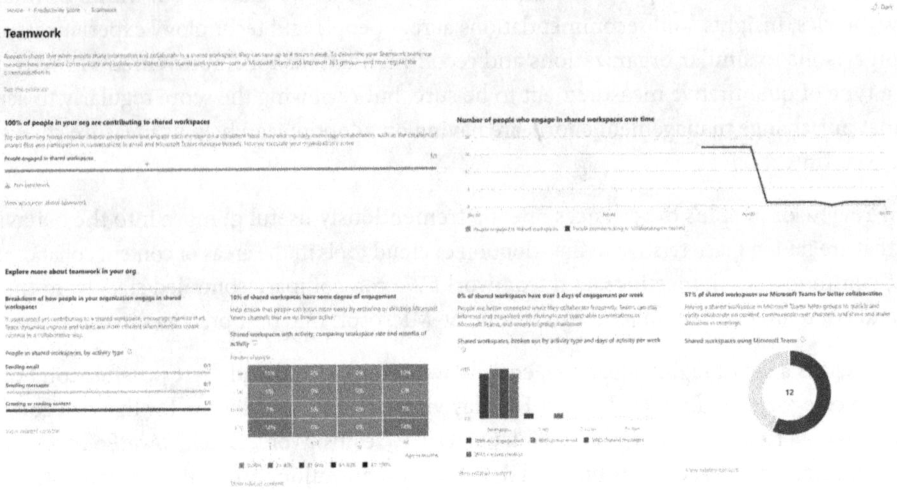

Figure 10.4 – Viewing the Teamwork experience details from the Productivity Score page

Viva Insights is another tool that can provide useful information to gauge adoption and employee engagement with M365 tools. It can show personal insights, which are also emailed daily, but the **My team** and **My organization** tabs can provide managers, team leaders, and executives information on the habits of their group relative to meetings and other forms of collaboration. In addition, insights related to how much time employees spend working after hours can indicate a source of change fatigue.

Here is an example of the **Accelerate change** report from Viva Insights, showing how influencers in the company can help drive change with a much larger percentage of the workforce if they become active champions of those change efforts:

Figure 10.5 – An organization report from Viva Insights

Qualitative – sentiment analysis

Understanding the feelings and sentiments toward a tool such as an intranet or the process of adopting it is vital to its long-term success. However, it is more difficult to measure it with accuracy. If opinions are generally negative, we either have not communicated the value well, or something needs to change in our design or technical approach.

If opinions are generally positive, we need to fully understand what's working well so that we can reinforce those elements and keep them fresh so that the positivity doesn't wane too quickly. The questions we should ask here are as follows:

- Do we have direct feedback from survey responses completed by our employees?

- What is the perception at each level of the organization – front-line workers, office staff, information workers, middle managers, leadership, and executives?

- Are we tracking sentiment using available tools or automation to scan messages?

- Do those who are sponsoring the initiatives feel they are getting their money's worth?

Some tools we can use here are as follows:

- **Surveys**: This could be a great opportunity to leverage Microsoft Forms to create and present survey questions that seek to gather overall sentiment and specific feedback on what's working well and where improvement is needed.

- **AI to mine sentiment and opinion**: If we don't think asking for direct feedback is working or is not giving us the full picture, we can rely on AI to help gather sentiment from emails, Teams chats, or other sources of commentary. The **Azure Cognitive Service for Language** is a set of **machine learning** (**ML**) and **artificial intelligence** (**AI**) tools that can help us provide labels on negative, neutral, or positive comments. Opinion mining can help users define subjects, opinions on those subjects, and the nature of the sentiment for each. More information on this service can be found here: `https://docs.microsoft.com/en-us/azure/cognitive-services/language-service/sentiment-opinion-mining/overview`. We should take extra care to inform our users of these services and let them know their privacy is important and protected.

Each successful change initiative may have unique criteria, but all successful adoption and change management efforts are ones where we've built awareness of the change, built desire for the change and how it can solve pain points, bolstered knowledge through a variety of learning methods, developed ability over time with coaching and support, and driven reinforcement through rewarding good behaviors and continual improvement. This five-step approach to personal change is what we'll explore next.

Implementing successful change with ProSci ADKAR

Some activities support organizational change and technology adoption that just make sense – training people, making sure business requirements are met by technical solutions, and communicating to our user base. However, even though they are common sense, the best practices are not always followed, nor are the activities owned by someone who is empowered to see them through. The role of the change manager should be seen as a discipline unto itself.

It is informed by industry standards and extensive research, and there are multiple methodologies and tools available to make change successful.

There are many change management models and frameworks to choose from. The Association of Change Management Professionals provides a CCMP certification. The Kübler-Ross model, most commonly associated with human grief, has been applied to how employees react to change, which can help with resistance management.

As noted previously, the ProSci model has become widely adopted and is tightly integrated with Microsoft. The model is known as **ADKAR**, an acronym for **awareness, desire, knowledge, ability, and reinforcement**.

The ADKAR process is linear, so whichever stage is the most lacking for an organization is where the attention should be focused before moving on. So, the training practice within your organization may be strong, but people may not understand why a change (such as a new corporate intranet) is being introduced. So, the focus would be on building awareness and desire before making any attempts to train employees on how to use or maintain that intranet.

The ADKAR model can be used to consistently drive change initiatives in your organization.

The model is best employed when the **ProSci Change Triangle** (**PCT**) model reveals the big picture of how ready the organization is for the change. The PCT reveals where the change objectives, leadership and sponsorship, project managers, and change managers are aligned or misaligned. With these four aspects aligned, we can practically execute the ADKAR model with activities that drive adoption, as defined here:

- **Awareness**: Build the why with communications sent by email, Teams chat, posters in the lunchroom, or whatever media works for your organization. If it's possible to ask for input on the potential change in advance, all the better, as communications should happen as early in the process as possible.

- **Desire**: Focus on WIIFM with specific business scenarios and pain points that will be addressed by the change. Begin to recruit champions and build the vision for the change and the stories to tell. We can hold exciting days and ask executives to record videos to kick things off.

- **Knowledge**: Define the training needs for the identified personas so that you can include guides and documentation, live training, and self-service training needs. Hold recurring Q&A sessions to address questions and supplement other activities.

- **Ability**: Building knowledge into repeatable ability takes time and continued care. Attending training doesn't mean you can immediately be a productive user. Sending tips and tricks emails, soliciting feedback, and having a collaborative space for people to support one another increases ability.

- **Reinforcement**: An extension of ability over time, refreshed and renewed.

Adoption reports can help us understand where to focus our efforts to reinforce change as time passes. Offering recognition in the form of praise, gamification, image badging, and so forth can help sustain the initial desire and deepen adoption.

If we take an example from the COVID pandemic, many organizations had to quickly shift to a hybrid or remote work model. As a result, organizations either started using or more fully adopted Microsoft Teams. To keep people connected, we may start using Teams meetings. To process this change using the ADKAR model, we must follow ADKAR's principles:

- **Awareness**: We needed to let everyone know the whys. The pandemic may have been the driver. The organization may have already owned licensing for M365 but may have not used Teams meetings. To get people connected quickly without additional cost may have been the rationale.

- **Desire**: We need to get people on board with the change. Each set of personas may have had different needs. Communicating how easy it is to host a meeting from anywhere on any device or being able to easily share meeting recordings with people inside and outside the organization may have driven the desire.

- **Knowledge**: Offering live training sessions on the life cycle of a meeting, along with one-page cheat sheets for how to use the functions within a meeting, may help people feel better prepared for the change of moving to virtual meetings.

- **Ability**: Microsoft updates the Teams client every couple of weeks, so keeping people apprised of the newest features helps them become more proficient. Perhaps tentative users who become comfortable with the tool can become a champion to help others.

- **Reinforcement**: Providing ongoing tips and tricks communications, along with surveys asking employees how they are adapting and eliciting their suggestions, can reinforce the desire and knowledge. Reminders from leadership acknowledging the struggles of working remotely but praising the success of adoption may have also reinforced the reasons for the change.

One of the best ways to start, fuel, and tend to change over time is making sure that people readily have the information they need to learn and grow.

Sustaining successful change with self-service learning

Successful adoption of new technologies starts with awareness of what's coming and the building of desire. To me, the rest of the ADKAR model focuses on thoroughly understanding the how-to aspects of the change being adopted. When that process of education and knowledge transfer begins, there is often a focus on live instructor-led training. While these are great opportunities to share useful information in a short time and a chance to ask questions, the amount of information retained is pretty small, especially if users of the tool are delayed.

To achieve ability and facilitate reinforcement, there are two key factors in my experience. The first is support from peers to help one another learn in the context of their daily work. This is why champions are so important. The other factor is having meaningful training material available to reference when it's needed. There is certainly no shortage of third-party training options available, but there are two options available from Microsoft that can be quickly implemented to assist with the ongoing adoption of SharePoint Online and other cloud tools – Viva Learning and the learning pathways site template.

Viva Learning

This tool is free with your SharePoint licensing, though it can include additional paid sources of training. Viva Learning is a Teams app that includes free content from LinkedIn Learning, Microsoft Learn, and Microsoft 365 Training by default. The latter options provide us with a very useful set of resources about the collaboration and communication tools within M365. Even if we don't leverage any custom learning content, this more than justifies the deployment of the app, in my opinion.

Someone in the knowledge admin or M365 admin role can manage the content sources. A Teams admin will be needed to deploy the app. People can choose their interests or can recommend content to each other in Viva Learning, which means that champions or change managers may be able to direct courses to employees over time, or they can support one another.

We can also configure SharePoint as a learning content source to make our content part of the Viva Learning experience.

The knowledge admin can work with the SharePoint admin to define a site where content will live. This is a list called the Learning App Content Repository.

Here, links can be set to folders that contain training content, assuming it is either Word, Excel, PowerPoint, PDF, a `.m4a` audio file, or a `.mov`, `.mp4`, or `.avi` video file. The Learning service uses the URLs we provide to generate metadata or to update it as content changes, with the refresh taking up to 24 hours.

M365 Learning Pathways

This is a solution that is found in the SharePoint LookBook, which can be downloaded from `https://lookbook.microsoft.com/details/3df8bd55-b872-4c9d-88e3-6b2f05344239`. The site has starter content plugged into a handful of pages:

Figure 10.6 – The Microsoft 365 learning pathways home page

While this is still a viable solution, it does predate Viva Learning, and the content may feel less robust. A tenant admin can deploy the solution from the link provided earlier.

It essentially creates a communication site supported by a SharePoint app (deployed to the tenant app catalog) that retrieves content from Microsoft. Since this is just a SharePoint site at its core, it can be branded like any other site to fit your organization's needs. You can either use the content provided or add your own training playlists.

This solution can coexist with Viva Learning if needed. If you're starting from scratch with these tools, Viva Learning would be my recommendation as the best place to start. If you have already invested time and content with your own learning pathways site, it can stand separately from Viva Learning, or we can look to integrate the two. Viva Learning does not know about or use the structure defined in learning pathways, but any custom content you add to libraries on the learning pathways site can be added to the Learning App Content Repository made available by Viva Learning.

Summary and planning document

In this chapter, we've explored aspects of a SharePoint Online deployment that focuses less on technical configuration and development and more on the other factors that lead to successful adoption. Managing the change purposefully and making sure that people understand the value and usage of our tools is a great return on our investment.

Here are some questions to ask as we add adoption and change management to our planning document:

- Does the organization have personnel trained in change management?
- What opportunities exist to train the organization on the adoption of best practices and the ADKAR model?
- Is there a defined set of technical change champions? If not, can a message be crafted and sent from management to recruit them?
- Are the pain points and business values clearly defined for the SharePoint Online solution we are planning?
- Does the organization have internal training resources?
- Have people been identified for the knowledge manager role if they're using Viva Learning?
- Does the project management office know about change management principles?
- Are middle managers being made aware of changes that affect them as well as changes that will affect their direct reports?
- What are the most effective training and communication tools and media in your organization?
- Is Productivity Score enabled in your admin center?
- Have site owners been trained on how to view usage stats for their sites?

It's appropriate that we close on adoption because that is the long-term goal of the planning efforts that we've undertaken together. Making a solution our own, embracing, it and becoming a champion of it creates satisfaction for us and our users. I hope you've found lots of useful information in these pages that you can use to build your plan to achieve more with SharePoint Online and all the ever-improving collaboration tools in Microsoft 365. Thanks so much for reading!

Index

D

S

V

W

Y

Packt.com

Subscribe to our online digital library for full access to over 7,000 books and videos, as well as industry leading tools to help you plan your personal development and advance your career. For more information, please visit our website.

Why subscribe?

- Spend less time learning and more time coding with practical eBooks and Videos from over 4,000 industry professionals

- Improve your learning with Skill Plans built especially for you

- Get a free eBook or video every month

- Fully searchable for easy access to vital information

- Copy and paste, print, and bookmark content

Did you know that Packt offers eBook versions of every book published, with PDF and ePub files available? You can upgrade to the eBook version at packt.com and as a print book customer, you are entitled to a discount on the eBook copy. Get in touch with us at customercare@packtpub.com for more details.

At www.packt.com, you can also read a collection of free technical articles, sign up for a range of free newsletters, and receive exclusive discounts and offers on Packt books and eBooks.

Other Books You May Enjoy

If you enjoyed this book, you may be interested in these other books by Packt:

Microsoft 365 and SharePoint Online Cookbook

Gaurav Mahajan, Sudeep Ghatak

ISBN: 9781838646677

- Get to grips with a wide range of apps and cloud services in Microsoft 365

- Discover how to use SharePoint Online to create and manage content

- Store and share documents using SharePoint Online

- Improve your search experience with Microsoft Search

- Leverage the Power Platform to build business solutions with Power Automate, Power Apps, Power BI, and Power Virtual Agents

- Enhance native capabilities in SharePoint and Teams using the SPFx framework

- Use Microsoft Teams to collaborate with colleagues or external users

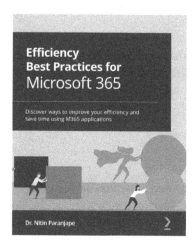

Efficiency Best Practices for Microsoft 365

Dr. Nitin Paranjape

ISBN: 9781801072267

- Understand how different MS 365 tools, such as Office desktop, Teams, Power BI, Lists, and OneDrive, can increase work efficiency

- Identify time-consuming processes and understand how to work through them more efficiently

- Create professional documents quickly with minimal effort

- Work across multiple teams, meetings, and projects without email overload

- Automate mundane, repetitive, and time-consuming manual work

- Manage work, delegation, execution, and project management

Packt is searching for authors like you

If you're interested in becoming an author for Packt, please visit `authors.packtpub.com` and apply today. We have worked with thousands of developers and tech professionals, just like you, to help them share their insight with the global tech community. You can make a general application, apply for a specific hot topic that we are recruiting an author for, or submit your own idea.

Share Your Thoughts

Now you've finished *SharePoint Architect's Planning Guide*, we'd love to hear your thoughts! Scan the QR code below to go straight to the Amazon review page for this book and share your feedback or leave a review on the site that you purchased it from.

`https://packt.link/r/1803249366`

Your review is important to us and the tech community and will help us make sure we're delivering excellent quality content.